UMOP 31

University of Massachusetts Occasional Papers 31:

Papers in Experimental Phonetics and Phonology

Edited by
Kathryn Flack and Shigeto Kawahara

Reproduced and distributed by

GLSA
(Graduate Linguistics Students' Association)
Department of Linguistics
South College
University of Massachusetts
Amherst, MA 01003-7130
U.S.A.

glsa@linguist.umass.edu
http://www.umass.edu/linguist/GLSA

Cover Design: Teruyuki Kawamura

ISBN: 1-4196-0681-6

Preface

Kathryn Flack and Shigeto Kawahara, editors

Much recent phonological work at the University of Massachusetts addresses formal questions about grammars through experimental methods. This volume presents a collection of recent papers by UMass students and faculty on this theme.

The papers can be broadly classified into two types of investigation. Papers focusing on phonological acquisition test whether observations based on adult language behavior and informed by our hypotheses about the structure of a phonological grammar can accurately predict language learning behavior. Pater and Tessier examine whether phonotactic knowledge in an adult's native language informs their learning of a novel language. Chambless examines child speech, testing the prediction that the range of word-medial syllabification possibilities allowed by a pre-adult OT grammar accurately predicts various context-dependent patterns of cluster simplification.

Many papers also address the ways and extent to which phonetics informs phonology, as well as the influence of phonology on phonetics. Flack tests the hypothesis that acoustic properties of laterals determine their phonotactic distribution in Australian languages. Kawahara notes that a voicing contrast in Japanese geminates is more easily neutralized than is the voicing contrast between singletons, and investigates whether this asymmetry follows from perceptual factors. Gelbart uses a collection of phonetic properties of French loanwords in English to investigate whether the presence of unambiguously "French" phonetic cues can bias a listener to perceive other ambiguous cues present in the stimulus as "French" as well, thus using shifts in perceptual boundaries as evidence of a more complex structure within the phonological lexicon. Shinya investigates whether phonetic perception of English epenthetic stops is shaped by phonological restrictions and/or lexical frequency. Kingston reports the results of two experiments that use prosodic manipulations to determine whether the effects of vowel height and obstruent voicing contrasts on F0 are automatic or controlled.

Table of contents

Phonotactics and alternations:
Testing the connection with artificial language learning[*]

Joe Pater and Anne-Michelle Tessier

University of Massachusetts, Amherst

ABSTRACT

Phonological alternations often serve to modify forms so that they respect a phonotactic restriction that applies across the words of a language. Since Chomsky and Halle (1968), it has been assumed that an adequate theory of phonology should capture the connection between phonotactics and alternations by deriving them using a shared mechanism. There is, however, no psycholinguistic evidence that speakers actually do use a single mechanism to encode phonotactics and alternations. In this study, we used an artificial language learning experiment to test whether an alternation that meets a phonotactic target is easier to learn than one that does not. The two groups of adult subjects learned mini-languages that differed only in whether the alternations were motivated by the phonotactics of their native language. The results suggest that phonotactic knowledge does aid in the acquisition of alternations, and also provide a novel example of the influence of the first language on second language learning.

1. Introduction

The phonotactics of a language restrict its set of possible sound sequences. One phonotactic generalization of English is that words cannot end with a sequence of a

---

[*] This research has been funded by NICHD/NIH grant 2T32-HD-07327-06A1. Thanks to John Kingston for his help in preparing the experiment, and to him as well as Angela Carpenter, Della Chambless, John McCarthy, Lisa Selkirk, Joe Stemberger, Janet Werker, Colin Wilson, and the participants in the Workshop on Markedness in the Lexicon at MIT and ICPhS15 for useful discussion.
[2] In fact, the restriction is broader: tautosyllabic obstruent clusters must share a single voicing specification (i.e. [+voice][-voice] sequences are banned too).

Kathryn Flack and Shigeto Kawahara (eds.), UMOP 31, 1-16.

voiceless obstruent followed by a voiced one (see (1a) below)[2]; another is that English monosyllabic words cannot have a rime which consists only of a lax (short) vowel, as in (1b):

(1)   a. *No [-voice][+voice] obstruent sequences:*
          *[bæpz] *[fowtz] *[hikd]
         cf. [bæps] [fowts] [hikt]

     b. *No monosyllabic CV words:*
         [blɪ],[gɛ],*[flʊ]
        cf. CVV: [blij], [gej], [fluw]; CVC: [blɪt], [gɛk], [flʊd]

The phonological *alternations* of a language are the changes in the shape of morphemes that depend on their phonological context. In English, the plural and past tense morphemes show a voicing alternation depending on the last consonant of the root, as illustrated in (2):

(2)   a. *Plural alternation between [z] and [s]*
          dog[z], job[z], rod[z]   *vs.*   dock[s], mop[s], lot[s]

     b. *Past tense alternation between [d] and [t]*
         rob[d], log[d], buzz[d]   *vs.*   dock[t], lop[t], miss[t]

Another English alternation holds between [k] and [s] in front of certain suffixes:

(3)   *Velar [k]at the end of Latinate stems: Alternating with [k] before -ity, -ism:*
    electri[k], lyri[k], romanti[k]   electri[s]ity, lyri[s]ism, romanti[s]ism

       The phonotactic generalizations in a language are often connected to its alternations in that the alternations enforce phonotactic well-formedness by eliminating structures that are also absent from the language's lexicon. For example, the voicing alternation in (2) ensures that plurals and past tense forms of English conform to the general phonotactic restriction in (1a).

       This connection between alternations and static phonotactics was recognized by Chomsky and Halle (1968):

(4)   ...regularities are observed within lexical items as well as across certain boundaries – the rule governing voicing of obstruent sequences in Russian, for example – and to avoid duplication of such rules in the grammar it is necessary to regard them not as redundancy rules but as phonological rules that also happen to apply internally to a lexical item. (p. 382)

Subsequent research cast doubt on the viability of a purely rule-based approach to the duplication problem, and this became one of the early arguments for the introduction of constraints into phonological theory (see especially Kenstowicz and Kisseberth 1977, 1979).

In Optimality Theory (OT; Prince and Smolensky 1993/2004), the duplication problem is solved by deriving phonotactics and alternations from a single set of constraints (McCarthy 2002; Hayes 2004). An OT grammar of English assumes a ranking like the one in (5) below, where a Markedness constraint AGREE[VOICE], which requires that adjacent obstruents match for voicing, ranks above the Faithfulness constraints against changing input voicing:

(5)   *An English ranking*
       AGREE[VOICE] >> FAITH[VOICE]

This ranking will enforce the English phonotactic restriction in (1a). Static phonotactics are dealt with in OT by having the grammar act as a filter. The grammar of a language should be able to take any underlying form as an input, and yield an output that conforms to the languages' phonotactics (Richness of the Base, Prince and Smolensky 1993). In the case at hand, any underlying input with a [-voice][+voice] obstruent sequence, like /boopz/ in (6) below, will be transformed to one with matching voicing (6b), which satisfies AGREE[VOICE]. The faithful candidate (6a) is ruled out because FAITH[VOICE] ranks beneath AGREE[VOICE]:

(6)

/bæpz/	AGREE[VOICE]	FAITH[VOICE]
a. bæpz	*!	
b. ☞ bæps		*

In addition, the same ranking will drive the alternation between the plural allomorphs [z] ~ [s]. Given a voiceless-final stem and a plural morpheme /z/, AGREE[VOICE] will again rule out the mismatched candidate (7a), and choose voicing assimilation for the winner (7b):[3]

(7)

/bʊk/ + /z/	AGREE[VOICE]	FAITH[VOICE]
a. bʊkz	*!	
b. ☞ bʊks		*

2.   The connection between phonotactics and alternations in acquisition

While there is much agreement amongst phonologists that an adequate theory of phonology should deal with the duplication problem, it remains an open issue whether language learners – either of first or second languages – do use a unified mechanism to encode phonotactics and to generate alternations.

---

[3] For a recent OT treatment of English word-final alternations in voicing, see Bakovic (2004).

One reason that one might be doubtful about the existence of a tight link between phonotactics and alternation is due to the fact that there are alternations that serve no phonotactic aim. In English, velar softening produces an alternation between final [k] ~ [s], as in (8a) below. However, this alternation does not seek to match any phonotactic restriction of English. It is not the case that underlying /k/ is becoming [s] due to a phonotactic demand, since [k] is allowed in this environment (e.g. ric[k]ety).

(8)   a. *Velar softening alternation...*   b. *... with no phonotactic motivation*
           electri[k] vs. electri[s]ity   c.f. ric[k]ety, cate[k]ism
           lyri[k] vs. lyri[s]ism

Pierrehumbert (to appear) presents *wug*-test results showing that English speakers do have productive knowledge of velar softening. This means that they did learn the alternation, even without phonotactic support. By extension, it could be argued that all alternations are learned from raw observation of changes in the phonological structure of morphemes, without reference to any knowledge of related lexical regularities.

Nonetheless, "conspiracies" between phonotactics and alternations pervade the phonologies of the world's languages. Based on this observation, one might hold that phonotactically-based alternations enjoy a special status, as in the single-mechanism account detailed above. In the realm of acquisition, a resulting prediction is that phonotactic knowledge should aid in the learning of alternations. This is explicitly claimed to be the case in recent L1 learnability work in Optimality Theory (Hayes 2004, Prince and Tesar 2004, Tesar and Prince 2004), in which phonotactic learning precedes, and informs, the learning of alternations.

While much is known about the acquisition of phonotactics, in both perception and production, comparatively little is known about the acquisition of alternations (although see Berko 1958; MacWhinney 1978; Derwing and Baker 1986; Bernhardt & Stemberger 1998; Kerkhoff to appear.) With respect to the current question of their interaction, as far as we know there is no empirical evidence. This lack of empirical results is not surprising, due to the difficulty of finding a testing ground in natural language. It is unlikely that naturalistic language acquisition will ever afford the opportunity to compare the ease of acquisition of two alternations that differ only in whether they are phonotactically motivated.

In this study, we set out to bypass this problem by studying the acquisition of an artificial language. For other applications of artificial language learning to the study of phonological acquisition, see e.g. Esper (1925), Schane *et al.* (1974), Bybee and Newman (1994), Curtin *et al.* (1998), Dell *et al.* (2000), Saffran (2001), Goldrick (2002), and Pater (2003), as well as Nowak et al. (2003), Wilson (2003) and Carpenter (to appear), discussed in section 4.

In our study, we compare the learning of two languages: one with a phonotactically motivated alternation, and one with a non-phonotactically motivated alternation. The phonotactic motivation comes from the native language of the subjects, who were adult native speakers of English. We assume that second language acquisition involves creation of a new grammar, using the same resources as first language acquisition (though other cognitive strategies may be used as well). One major difference, however, is that the initial state of second language acquisition is the final state of first language acquisition, and thus that phonotactic properties of the subjects' native language are available to help in learning the artificial languages. Our research question is thus:

(9)   Does an L1 phonotactic restriction aid in the acquisition of an L2 alternation?

It is worth emphasizing that our assumption about the transfer of L1 phonotactic knowledge into L2 is just that: an assumption. While it is clear that phonotactic restrictions affect L2 learners' productions (for example, in triggering epenthesis in consonant clusters), we cannot be sure that they would transfer at a more abstract level (for example, in phonotactic well-formedness judgments, or in the task below). Therefore, a positive answer to our research question would provide novel evidence of transfer in L2 learning.

2.1.   Studying the acquisition of alternations with artificial language learning

The English phonotactic restriction our artificial language exploits is the Minimal Word restriction from (1b): monosyllables that end in a lax vowel are ill-formed (e.g. *[blɛ]). Data from Moreton (1999) provides evidence that English speakers do have productive knowledge of this restriction. In that study, listeners were more likely to identify a vowel that is ambiguous between [ij] and [ɪ] as [ij] in the word-final context than in a context where both are permitted. In addition, Cebrian (2002) shows that native English speakers, and Catalan learners of English, use this restriction in interpreting the morphological composition of nonce words. Since English provides no alternations to repair subminimal words, we thus addressed our research question by asking whether English speakers are able to learn such an alternation more readily than a comparable one without a phonotactic purpose.

Cross-linguistically, alternations can augment sub-minimal words in a variety of ways – for example, by lengthening or epenthesizing a vowel (McCarthy & Prince 1991). We chose consonant epenthesis, as in Cupeño (Crowhurst 1994, Lombardi 2003), since vowel lengthening does occur elsewhere in English. The cross-linguistically most common epenthetic consonant is likely to be glottal stop; however because [ʔ] is difficult to perceive, we chose the voiceless alveolar [t] instead, as used in Axininca Campa (Payne 1981, McCarthy and Prince 1993).

Both languages that we constructed contain singular and plural words, in which the plural is marked with the suffix /-so/. In Language 1, epenthesis is used to avoid words that would be sub-minimal in English. It applies to avoid word-final lax vowels, as in the singulars in (10a-c), but not if the singular ends in a tense vowel (10d), or a consonant (10e):

(10)   *Language 1*

	Root	Plural	Singular
a.	/blɪ/	[blɪso]	[blɪt]
b.	/gɛ/	[gɛso]	[gɛt]
c.	/flʌ/	[flʌso]	[flʌt]
d.	/blej/	[blejso]	[blej]
e.	/glɛk/	[glɛkso]	[glɛk]

In the other language, epenthesis applies in a similar fashion, but after front, rather than lax vowels (11a-c).[4] It does not apply after back vowels (11d), or after consonants (11e):

(11)   *Language 2*

	Root	Plural	Singular
a.	/lij/	[lijso]	[lijt]
b.	/blej/	[blejso]	[blejt]
c.	/træ/	[træso]	[træt]
d.	/fuw/	[fuwso]	[fuwt]
e.	/gluwk/	[gluwkso]	[gluwk]

If alternations are learned simply by pattern recognition and formalization (by e.g. rule or constraint) – without reference to other phonological knowledge – these two languages should be equally easy to learn. This is highlighted by the similarity between the following rules describing the alternations:

(13)   Language 1:   Ø → t /   V ___ #
                              [-tense]

       Language 2:   Ø → t /   V ___ #
                              [-back]

However, if English learners draw on their knowledge of the phonotactics of their native language, then language 1 should be easier to learn.

At this point, the reader may concerned by the fact that these two languages differ along another dimension, namely natural language attestedness. It is clear that languages like Language 1 which repair Minimal Word violations are fairly common, whereas Language 2's 'repair' of final front vowels is probably unattested. Clearly, it is of interest to what extent a positive result in our experiment can be attributed to each of these two factors individually; however, it seems very likely to us that the more powerful boost to the acquisition of Language 1 over Language 2 for our learners would be the phonotactic

---

[4] For 6 of 14 subjects, epenthesis was triggered by back, rather than front vowels; the scores of these two groups do not differ significantly.

motivation supplied by English. In section 4, we will return to this question of attestedness, and discuss the teasing apart of multiple possible influences on artificial language learning in general.

In terms of Optimality Theory, English-speaking learners of Language 1 will only need to establish a ranking of faithfulness constraints that would choose epenthesis as a repair. The ranking of the relevant markedness constraint(s) (e.g. FOOTBINARITY; Prince and Smolensky 1993) above faithfulness constraints will already be in place, assuming an initial Markedness >> Faithfulness ranking bias that will never have been disturbed because English words do respect the ban on subminimality. On the M >> F learning bias, see especially Smolensky 1996; Hayes 2004; Prince and Tesar 2004; see also Ota 2001 for evidence of child language word minimality effects in Japanese, in which the adult language permits sub-minimal words.

Learners of Language 2 will be faced with a more complex task. The exact nature of that task depends on how "unnatural rules" are learned: as just discussed above, epenthesis following front vowels seems to be unattested cross-linguistically. Under one account (Hayes 1999), they will have to construct a language-specific constraint to drive the alternation, and then fix its ranking with respect to the rest of the hierarchy, as well as establishing the ranking of faithfulness constraints.

3. The experiment

3.1. Materials

Three sets of plural-singular pairs (nonce words in English) were created for each language. In each language, two sets were composed of vowel-final roots, which either induced alternations, or not. For each of the vowel categories (i.e. tense, lax, back, front) there were three vowels, with four instances of each. The other set consisted of consonant-final roots. There were twelve pairs for each set; the following are representative examples:

(14) *Language 1*

V-Final roots (Alternating)	*V-Final roots (Non-alternating)*	*C-Final roots*
[kɛso] [kɛt]	[blejso] [blej]	[trɛtso] [trɛt]
[glɪso] [glɪt]	[lijso] [lij]	[vejtso] [vejt]
[yʌso] [yʌt]	[pluwso] [pluw]	[vijkso] [vijk]

(15) *Language 2*

V-Final roots (Alternating)	*V-Final roots (Non-alternating)*	*C-Final roots*
[lijso] [lijt]	[vuwso] [vuw]	[ruwkso] [ruwk]
[blejso] [blejt]	[trowso] [trow]	[dijtso] [dijt]
[træso] [træt]	[vɑso] [vɑ]	[vijkso] [vijk]

The words were spoken by a trained phonetician in carrier phrases, and edited out for presentation via computer over headphones. Words were paired with picturable nouns (e.g. airplanes/airplane, trees/tree, balls/ball).

3.2. Subjects

Subjects were native speakers of English, with no knowledge of a second language beyond high school level. They were recruited by advertisement and word of mouth, and paid for participation. A between-subjects design was used; each subject learned one of the languages.

3.3. Testing and training

Throughout the experiment, subjects were seated in front of a computer screen, with headphones on. The experiment consisted of three phases, each containing a training block and a testing block with a randomized order of items. In training, plural/singular pairs were presented in turn, by displaying the visual referents on the computer screen and simultaneously playing the aural label over the headphones. Subjects always heard the plural first, and then the singular. Subjects pressed a key to move on to the next pair.

For the testing component, we followed Saffran *et al.* (1996) in using a forced choice task. In each test trial, subjects heard a plural word with its associated plural picture, followed by two possible singular forms of that word with the corresponding singular picture. An example test trial is illustrated in (16):

(16)   *Example test trial*

	X	A	B
	X	A	B
Audio:	[blejso]	[blej]	[blejt]
Visual:	*(picture of*	*(picture of*	*(picture of*
	apples)	*apple)*	*apple)*

By pressing a key, subjects had to choose between A and B as singular forms for X. Choices always differed in the presence of the final consonant. For the example in (16), the correct answer for learners of Language 1 would be A, while for Language 2 it would be B.

In each of the first two phases, subjects were first trained and then tested on 9 pairs, with each pair appearing three times in training, and twice in testing. In the third phase, the 18 now-familiar pairs were played once more in training mode for review. Then, subjects were tested on those 18 plus another 18 novel pairs, resulting in 36 total, again with each pair appearing twice.

Thus, subjects were only trained on half of the items, but eventually tested on all of them. The 'novel' test items in the third block allowed us to examine whether subjects had acquired the generalization, rather than having simply memorized the correct singulars for each trained plural.

3.4. Results

Data was collected from 29 subjects – 14 in Language 1 and 15 in Language 2. We report the results from only non-outliers (those within two standard deviations of the mean), leaving 11 in Language 1 and 14 in Language 2. Table (17) shows the mean number of correct responses across all subjects in the final testing block, separated into responses for the trained and novel items (standard deviations are given in brackets):

(17)   *Performance on all trained and novel items in final test block*

	Language 1 (n=11)	**Language 2** (n=14)
<u>Trained items</u>	0.94 (0.05)	0.96 (0.06)
<u>Novel items</u>	0.74 (0.06)	0.66 (0.14)

Both groups did well on the trained items; this may simply indicate that they were able to memorize the correct singular form. However on the novel items, the subjects learning the phonotactically motivated alternation of Language 1 also did better than those in Language 2. A single tailed t-test assuming unequal variance finds the between groups difference on the novel items to be significant (t(18) = 2.24, p < 0.02).[5]

These results provide support for the claim that L1 phonotactic knowledge does indeed play a role in the learning of L2 alternations. However, there is a way in which L1 phonotactic knowledge itself could have driven this pattern of results, without any reference to a new L2 phonology. In the Language 1 condition, subjects could have responded correctly to some of the test items merely on the basis of what is phonotactically allowed in English. For example, when faced with [gɛ] and [gɛt] as choices for the singular of [gɛso], the correct response [gɛt] is also the one that is well-formed in English. One might therefore speculate that the two groups did equally well in learning the alternation, but that learners of Language 1 responded correctly more often because they sometimes decided based on what was well-formed in their native language.

Under this alternative explanation, the learners of Language 1 should only outperform Language 2 on the alternating roots – those that trigger epenthesis in the singular. The non-alternating roots should be equally easy or difficult in both language conditions, since their test trials involve a choice between forms that are equally legitimate in English. To test this prediction of the alternative explanation, the following figure graphs the mean proportion correct for each of the two groups for the three root-types (see (14) and (15) for examples of each). Error bars indicate 95% confidence intervals.

---

[5] A t-test with all subjects, including outliers, has p < 0.01.

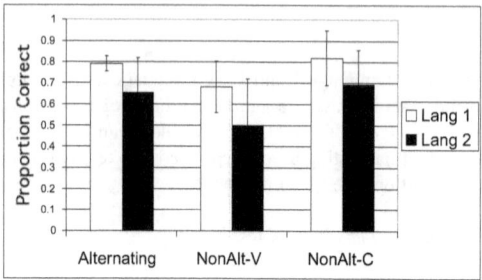

Figure 1. Performance on different root types.

As this graph indicates, subjects learning Language 1 did in fact do better on *all* root types. Poor performance on the non-alternating roots indicates that the learners were incapable of correctly determining the scope of the alternation. It seems that this was more of a problem for learners of Language 2 than Language 1, though this conclusion is still tentative, given that the between group differences within root types do not reach significance.

4.   Discussion

Overall, our results indicate that learners do not learn alternations by simply observing the changes to the phonological shape of morphemes, but rather are aided in this task by other aspects of phonological knowledge. This 'other knowledge', however, could come from more than one source, whose possibly independent effects we now consider.

4.1.   Potential influences on the acquisition of (L2) alternations

Thus far, we have mostly characterized the difference between the two languages in our experiment as a difference in L1 phonotactic motivation: Language 1's alternation prevents its singulars from violating an English restriction, while Language 2's alternation has no such explanation. However, it has already been briefly mentioned that our languages also differed in their similarity to the patterns of natural languages: as far as we know, no real language behaves like Language 2 in closing off open front-voweled syllables with an epenthetic consonant. Moreover, we know of no example of a phonotactic constraint which exclusively prohibits open front-voweled syllables, nor of any other alternation that is used to avoid them. Therefore, one alternative explanation for our results is that subjects had trouble with Language 2 because it contained an unattested, or "unnatural", phonological rule. This alternative explanation rests on the premise that learners can distinguish between possible and impossible rules – a premise that is clearly of no little interest.

Although we do not think that it played a role in our results, another possible influence on the acquisition of an alternation is its formal simplicity. Determining the simplicity of a alternation can depend heavily on theoretical assumptions of how the rules or constraints producing it are formalized, and is thus far from straightforward. However, as a starting point we can assume that rules that manipulate more features are more complicated (as in Chomsky and Halle's 1968 feature counting metric). As discussed in section 2, we attempted in our experiment to choose two alternations that were as similar as possible – differing only in whether the vowel feature [-tense] or [-back] triggered final [t] epenthesis.

With respect to the L2 acquisition of an alternation, one plausible hypothesis is that all the factors above can ease or inhibit the process. In this view, the spectrum of alternations and their relative learnability could be diagrammed as below (18):

(18)   Hypothesized spectrum of factors influencing an alternation's learnability

Easiest to learn   Hardest to learn

• *phonotactically motivated*   *phonotactically unmotivated* •
• *attested in natural language*   *unattested in natural language* •
• *formally simple*   *formally complex* •

Because our experiment was designed as an initial foray into the use of an artificial learning methodology for the study of alternations, we designed our materials to provide a clear contrast on the spectrum: pitting Language 1's natural rule with motivation in English phonotactics against Language 2's unnatural rule with no connection to English patterns.

To move on to the more subtle question of how each factor contributes to ease of acquisition, we must test for each factor's influence on learning in turn while keeping other factors constant. Designing such experiments can pose a considerable challenge. To test for phonotactic motivation independent of cross-linguistic attestedness or formal simplicity, one would have to find two nearly identical attested alternations that apply in almost exactly the same environment, and teach them to subjects whose native language provides only the phonotactic motivation for one of them, and not the alternation itself. Testing attestedness as the criterion factor is somewhat easier: this only requires that the nearly-identical rules are unmotivated by L1 phonotactics (though statistical tendencies in the L1 are less easy to control for; C. Wilson, p.c.). Recently, several studies have examined attestedness as a factor (or the related notion of 'naturalness'),[6] as well as formal simplicity, using artificial language learning paradigms. These studies have all found differences between experimental conditions, which bodes well for the use of artificial language learning in this domain. However, they also highlight the difficulties in teasing apart the factors at play.

Pycha *et al.* (2003) used a similar methodology to ours to compare the acquisition of a suffix whose vowel backness – [ʊ] vs. [ɛ] – was predictable in some way from the root vowel. In their Language 1, the suffix alternated so that it agreed (i.e. assimilated) with the preceding vowel in backness. In Language 2, it consistently disagreed with the preceding

---

[6] In general, the terms "natural" and "attested" can both be taken to mean "cross-linguistically attested". However, there are attested phonological alternations that are sometimes categorized as unnatural, because they lack phonetic motivation, have many exceptions, and/or are rare (see esp. Stampe 1969): English velar softening is one example (Pierrehumbert to appear).

vowel in backness (i.e. it dissimilated). In the third language, the suffix was also predictable based on the final vowel, but some stem vowels caused assimilation, and others caused dissimilation. The first two languages were meant to test naturalness as a factor; backness assimilation was labeled as natural due to phonetic factors. However, as Pycha et al. note, dissimilation of this type is attested cross-linguistically (in Ainu, Ito 1984 and Yucatec Maya, Kramer 1998), so it is not clear that it is truly unnatural. Therefore, there is little that can be concluded from the fact that no difference was observed between the first two languages. There was a large difference between the success of learners of the third language and the first two. Since the rule for the third language is so complex, Pycha *et al.*'s conclusion that this complexity was the source of difficulty is likely correct, though it is possible that attestedness played a role, since the third language is clearly unattested.

Wilson (2003) also reports an artificial learning study of this type, which did compare an attested and unattested alternation. The two languages he constructed also had an alternating suffix, [-la] vs. [-na], whose shape was predictable in some way from the featural make-up of the preceding stem's final consonant. In the language with an attested pattern, [na] appeared if there was a nasal as the onset of the previous syllable; otherwise the suffix was [la]. The unattested language had the appearance of [na] linked with the presence of a dorsal in the previous syllable. Wilson found that learners of the attested alternation performed much better on a test with novel forms than did the learners of the unattested alternation. He also replicated this result with a dissimilation pattern, which similarly produced better results on the novel word test than did the unattested alternation. However, it is not clear that the processes that are compared in Wilson's study can be considered equivalent in all respects beyond their attestedness. For learners of the nasal assimilation and dissimilation languages, it is only necessary to keep track of the co-occurrence of a single feature: [+/-nasal]. Learners of the random alternation, however, must notice the co-occurrence of [+nasal] or [-nasal] with [dorsal]. This could hamper the learners of the random alternation in two ways. First, they may be less likely to notice the generalization: as a reasonable analogy, a surveyor of passing traffic would more likely notice that every Volkswagen is followed by another VW than notice that every Buick is followed by a VW. Second, they might have more difficulty in formalizing the generalization: assimilation and dissimilation rules are often given a special status, either though use of alpha notation or feature spreading conventions (see Wilson 2003 for discussion of this second alternative). Thus, this study's results do not seem to provide a clear test of acquisition ease as a sole function of attestedness.

Current work in this vein by Carpenter (to appear) provides what seems to be a good test case for attestedness alone. Instead of alternations, her experiment compares two stress rules, both based in part on sonority. The attested rule stresses the leftmost low vowel, and if there are no low vowels, the leftmost high one. The unattested rule is exactly the same, except that it reverses the preference: leftmost high, else leftmost low. Carpenter finds that both English and Québocois French subjects are more successful on a novel word test when they are learning the attested, rather than the unattested language. One potential confound here is that the inherent intensity of low vowels would lead to their being more easily associated with stress; this is the likely source of the cross-linguistic pattern. This concern is mitigated, however, by the fact that the intensity of the vowels in the experiment was controlled.

These studies provide some indication that cross-linguistic attestedness may influence the acquisition of phonological alternations, though they also highlight the methodological difficulties in addressing these important research questions.

5.   Conclusions

Our experimental results indicate that two very similar phonological alternations differ in how easily they are acquired: epenthesis following lax vowels was learned more successfully than epenthesis following front vowels. These alternations differ in whether they are motivated by the phonotactics of the L1, and in whether they reflect cross-linguistically observed phonological patterns. These results could be explained by transfer of L1 phonotactic knowledge into the L2, by access to universal phonological principles, or as we suspect to be accurate, by a combination of the two. Future research should offer insight into the relative contribution of these factors. More generally, we offer this study as a contribution to a line of research that appears very promising: the use of artificial language learning to study otherwise intractable questions in both first and second phonological acquisition.

References

Bakovic, Eric. 2004. All or Nothing: Partial Identity Avoidance as Cooperative Interaction. Ms., University of California, San Diego. Available on the Rutgers Optimality Archive: ROA #671-0804.
Berko, Jean. 1958. The child's learning of English morphology. *Word* 14:150-177.
Bernhardt, Barbara and Joseph Stemberger. 1998. *Handbook of Phonological Development: From the Perspective of Constraint-based Nonlinear Phonology.* San Diego: Academic Press.
Brown, Cynthia. 1998. The role of the L1 gramamr in the L2 acquisition segmental structure. *Second Language Research* 14:136-193.
Bybee, Joan and Jean Newman. 1995. Are affixes more natural than stem changes? *Linguistics* 33:633-654.
Carpenter, Angela. To appear. Acquisition of a Natural vs. Unnatural Stress System. In the proceedings of BUCLD 29.
Cebrian, Juli. 2002. *Phonetic Similarity, Syllabification and Phonotactic Constraints in the Acquisition of a Second Language Contrast.* Ph.D. dissertation, University of Toronto.
Chomsky, Noam and Morris Halle. 1968. *The Sound Pattern of English.* New York: Harper and Row.
Clements, G. N. and Engin Sezer 1982. Vowel and consonant disharmony in Turkish. In H. van der Hulst and N. Smith [eds.] *The Structure of Phonological Representations,* Part 2, 213-55.
Curtin, Suzanne, Heather Goad, and Joseph Pater. 1998. Phonological transfer and levels of representation: The perceptual acquisition of Thai voice and aspiration by English and French speakers. *Second Language Research* 14:389-405.
Davidson, Lisa. 2003. Articulatory and perceptual influences on the production of

non-native consonant clusters. In M.J. Solé, D. Recasens, and J. Romero [eds.] *Proceedings of the 15th International Congress on Phonetic Sciences*, Barcelona.

Dell, Gary. S., Kristopher D. Reed, David R. Adams and Antje S. Meyer. 2000. Speech errors, phonotactic constraints, and implicit learning: A study of the role of experience in language production. *Journal of Experimental Psychology: Learning, Memory, and Cognition*, 26:1355-1367.

Derwing, Bruce, L. and W. J. Baker. 1986. Assessing morphological development. In P.J. Fletcher and M. Garman [eds.], *Language Acquisition: Studies in First Language Development* (2nd edition). Cambridge: Cambridge University Press, 326-338.

Dupoux, Emmanuel & Sharon Peperkamp. 2002. Fossil markers of language development: phonological 'deafnesses' in adult speech processing. In J. Durand and B. Laks [eds.] *Phonetics, Phonology, and Cognition*. Oxford: Oxford University Press, 168-190.

Eckman, Fred and Gregory Iverson. 1997. Structure preservation in interlanguage phonology. In S.J. Hannahs and M. Young-Scholten [eds.] *Focus on phonological acquisition*. Amsterdam: John Benjamins, 183-207.

Esper, E. 1925. A Technique for the Experimental Investigation of Associative Interference in Artificial Language Material. *Language Monographs* 1.

Goldrick, Matt. 2002. *Patterns in Sound, Patterns in Mind: Phonological Regularities in Speech Production*. Ph.D. dissertation, Johns Hopkins University.

Hayes, Bruce. 1999. Phonological restructuring in Yidiny and its theoretical consequences. In B. Hermans and M. van Oostendorp [eds.] *The Derivational Residue in Phonological Optimality Theory*. Amsterdam: John Benjamins, 175-205.

Hayes, Bruce. 2004. Phonological acquisition in Optimality Theory: The early stages. In R. Kager, J. Pater and W. Zonneveld [eds.] *Constraints in Phonological Acquisition*, Cambridge, UK: Cambridge University Press, in press.

Hyman, Larry. 1995. Nasal consonant harmony at a distance: The case of Yaka. *Studies in African Linguistics*. 24:5-30.

Ito, Junko. 1984. Melodic dissimilation in Ainu. *Linguistic Inquiry* 15:505-513.

Kaun, Abigail. 1995. *The Typology of Rounding Harmony: An Optimality Theoretic Approach*. Ph.D. dissertation, UCLA.

Kenstowicz, Michael. 1994. Sonority-driven stress. Ms., MIT. Available on the Rutgers Optimality Archive: ROA #33-1094.

Kenstowicz, Michael and Charles Kisseberth, 1977. *Topics in Phonological Theory*. New York: Academic Press.

Kenstowicz, Michael and Charles Kisseberth. 1979. *Generative Phonology*. New York: Academic Press.

Kerkhoff, Annemarie. To appear. Acquisition of voicing alternations. To appear in S. Baauw and N.J. van Kampen [eds.] *Proceedings of GALA 2003*.

Kraemer, Martin. 1998. *A Correspondence Approach to Vowel Harmony and Disharmony*. SFB 282 working paper #107, Heinrich-Heine Universitaet-Duesseldorf.

de Lacy, Paul. 2002. *The Formal Expression of Markedness*. Ph.D dissertation, UMass Amherst. Amherst: GLSA Publications.

Leather, Jonathan and Allan James. 1991. The acquisition of second language speech. *Studies in Second Language Acquisition* 13:305-341.

MacWhinney, Brian. 1978. *The Acquisition of Morphophonology*. Monographs of the Society for Research in Child Development; 43, no 1.

Major, Roy C. 1994. Current trends in interlanguage phonology. In M. Yavas [ed.] *First and Second Language Phonology*. San Diego: Singular Publishing, 181-204.

McCarthy, John J. 2002. *A Thematic Guide to Optimality Theory*. Cambridge, UK: Cambridge University Press.

McCarthy, John J. and Alan Prince. 1996. *Prosodic Morphology 1986*. Technical report, Rutgers University Center for Cognitive Science. New Brunswick, NJ: Rutgers University.

McCarthy, John J. and Alan Prince. 1993. *Prosodic Morphology I: Constraint Interaction and Satisfaction.* Technical report, Rutgers University Center for Cognitive Science. New Brunswick, NJ: Rutgers University.
Moreton, Elliott. 1999. Evidence for phonotactic grammar in speech perception. In *Proceedings of the 14th Annual International Congress of Phonetic Sciences*, San Francisco.
Odlin, Terence. 1989. *Language Transfer: Cross-linguistic Influence in Language Learning.* Cambridge: Cambridge University Press.
Ota, Mitsuhiko. 2001. Phonological theory and the development of prosodic structure: Evidence from child Japanese. *Annual Review of Language Acquisition* 1, 65-118.
Pater, Joe. 2003. The perceptual acquisition of Thai phonology by English speakers: Task and stimulus effects. *Second Language Research* 19:209-223.
Payne, D. 1981. *The Phonology and Morphology of Axininca Campa.* Dallas, TX: SIL Publications.
Pierrehumbert, Janet. To appear. An unnatural process. Paper presented at the Eighth Conference on Laboratory Phonology, New Haven, June 2002. [To appear in proceedings].
Prince, Alan and Paul Smolensky. 2004. *Optimality Theory: Constraint Interaction in Generative Grammar.* Blackwell. [Originally published as Technical report, Rutgers University Center for Cognitive Science. New Brunswick, NJ: Rutgers University, 1993.]
Prince, Alan and Bruce Tesar. 2004. Learning phonotactic distributions. In R. Kager, J. Pater and W. Zonneveld [eds.] *Constraints in Phonological Acquisition.* Cambridge, UK, Cambridge University Press.
Pycha, Anne, Pawel Nowak, Eurie Shin and Shin Shosted. 2003. Phonological rule-learning and its implications for a theory of vowel harmony. In G. Garding and M. Tsujimura [eds.] *Proceedings of the 22nd West Coast Conference on Formal Linguistics.* Somerville, MA: Cascadilla Press, 423-435.
Rubach, Jerzy. 1984. Rule typology and phonological interference. In S. Elliason [ed.] *Theoretical Issues in Contrastive Phonology.* Heidelberg: Julius Groos Verlag, 37-50.
Saffran. J.R. 2001. The use of predictive dependencies in language learning. *Journal of Memory and Language* 44:493-515.
Saffran, Jennifer R., Elissa. L. Newport and Richard N. Aslin. 1996. Word segmentation: The role of distributional cues. *Journal of Memory and Language*, 35:606-621.
Schane, Sanford A., Bernard Tranel and Harlan Lane. 1974. On the psychological reality of a natural rule of syllable structure. *Cognition* 3:351-358.
Smolensky, Paul. 1996. *The Initial State and "Richness of the Base" in Optimality Theory.* Technical report, Department of Cognitive Science, Johns Hopkins University.
Stampe, David. 1969. The acquisition of phonemic representation. *Proceedings of the Fifth Annual Chicago Linguistics Society,* 433-444.
Tesar, Bruce, and Alan Prince. 2004. Using phonotactics to learn phonological alternations. In *The Proceedings of CLS 39, Vol. II: The Panels.* Available on Rutgers Optimality Archive: ROA #620.
Wilson, Colin. 2003. Experimental investigation of phonological naturalness. In G. Garding and M. Tsujimura [eds.] *Proceedings of the 22nd West Coast Conference on Formal Linguistics.* Somerville, MA: Cascadilla Press, 533-546.

Joe Pater and Anne-Michelle Tessier

Department of Linguistics
South College
University of Massachusetts
Amherst, MA 01003

pater@linguist.umass.edu
tessier@linguist.umass.edu

Sonority, syllable weight, and contextual asymmetries
in cluster production and reduction

Della Chambless

University of Massachusetts, Amherst

1.   Introduction

Consonant cluster production has been the focus of many investigations in the field of phonological acquisition (e.g. Bernhardt & Stemberger, 1998; Gnanadesikan, 2004; Ohala, 1996). Previous studies of cluster acquisition have provided accounts both of the choice of consonant preserved in stages in which reduction pervades, and of the observed order in which different clusters are acquired. Most investigations have involved clusters in word-initial position, though a few have dealt with intervocalic consonant clusters (e.g. Barlow, 2004; Bernhardt & Stemberger, 1998; Ohala, 2004).

The intervocalic (i.e. word medial) context is interesting because the syllabification of consonants is more difficult to establish unequivocally. While it is often assumed that an intervocalic cluster which is permissible word initially, is syllabified as an onset intervocalically,[1] there is an abundance of psycholinguistic evidence from adult English which shows this generalization not to hold across the board, specifically in particular prosodic contexts. For example, results from syllabification tasks (Treiman & Zukowski, 1990) show that the tendency of an intervocalic /CC/ sequence to be heterosyllabic is greater for /sC/ clusters than for stop-liquid clusters, for clusters preceded by stressed syllables than for those preceded by unstressed syllables, and for those clusters preceded by a stressed lax vowel than those following a stressed tense vowel.

Sonority constraints have been invoked to account for specific patterns of word-edge cluster acquisition and reduction (Fikkert, 1994; Gnanadesikan, 2004; Ohala, 1996; Pater, 1997; Pater & Barlow, 2003; Prince & Smolensky, 1993). As such constraints (e.g. *Sonority Sequencing*, **Liquid-Onset*, **Stop-Coda*, etc.) are syllable based, the precise status of intervocalic consonants (as onsets, codas, or both) should affect patterns of

---

[1] This follows from the *Principle of Legality* and the *Onset Maximization Principle* (Kirk, 2001; Pulgram, 1970; Selkirk, 1982; Treiman et al 1992; Treiman & Danis, 1988; Treiman & Zukowsky, 1990).

Kathryn Flack and Shigeto Kawahara (eds.), UMOP 31, 17-35.

acquisition order and reduction in such an environment. Precisely because medial consonants differ from word edge consonants in the ambiguity of their syllabic affiliation, positional asymmetries in cluster acquisition are predicted. For example, an s-stop cluster is marked in initial position because it violates either *Sonority Sequencing* or **Adjunct*; however, in medial position, it may be split across two syllables: <u>word medial</u> s-stop clusters are thus predicted to possibly precede <u>word initial</u> s-stop clusters in acquisition.

The goal of this study is to test predicted asymmetries between initial and medial consonant cluster production, in addition to differences among the 'medial' group which depend on both stress and vowel quality. Section 2 reports on an experiment carried out to investigate order of acquisition of initial and medial clusters in English. Section 3 analyzes reduction data emerging from the same experiment.

2.   Order of acquisition

2.1.   Introduction

While early utterances in child speech are characterized by cluster reduction everywhere, assumptions of current phonological theory involving universal constraints and the acquisition process lead to predictions concerning intermediate stages of cluster acquisition in which some clusters are present and some are absent. In general, both *positional* and *segmental* asymmetries are expected.

For example, as suggested above, word-medial position differs from word-initial position in that intervocalic /CC/ clusters can be syllabified as part of the same or different syllables: $[VC_1.C_2V]$ or $[V.C_1C_2V]$. As a consequence, even if complex onsets are not present in the child's speech, as verified by reduction in word-initial position, the preservation of both consonants in intervocalic position should be possible, as long as codas are already legal in the child's grammar. Furthermore, some heterosyllabic clusters are better than others, as a consequence of sonority constraints belonging to the *Syllable Contact* family.

I set out to investigate here whether the cluster type (s-stop vs. stop-liquid), position (initial vs. medial) and the prosody of the previous syllable (stressed-lax, stressed-tense, unstressed) affect the patterns of reduction in the speech of 2-3 year olds acquiring English. Predictions follow from the interactions of constraints which are described in the following section.

2.2.   Constraint interactions

2.2.1.   Word-initial clusters

While many languages permit complex onsets, the sequences of consonants which make up such a cluster are restricted. Clusters containing *stop-liquid* (e.g. [#pl]) sequences are allowed in English, while *liquid-stop* sequences (e.g. [#lp]) are not. Some such restrictions are attributed to the *Sonority Sequencing Principle (SSP)* (Clements, 1990). The SSP militates against word-initial consonant sequences in which sonority drops across the onset. For languages such as English, Dutch, and Italian, which generally obey the SSP, but permit /s-stop/ clusters word initially, it is commonly assumed that the /s/ is extraprosodic in word initial position (e.g. Barlow, 2001; Fikkert, 1994 for Dutch; Green, 2003; Steriade, 1982 for Attic Greek).[2]

In child speech, a difference in acquisition order has been noted for the following two classes of clusters: *obstruent-liquid* and *sC* clusters. While most evidence points to the later acquisition of *sC* clusters, a few studies have found evidence for the opposite order of acquisition (see Barlow, 2001 for a review of the literature). The grammar of a child who produces stop-liquid clusters but not s-stop clusters may be a product of the constraint ranking in (1).

(1)   *SSP >> Faith >> *Complex*

/blu/	SSP	Faith	*Complex
a. ➜ blu			*
b. bu		*!	
/stu/	SSP	Faith	*Complex
c. stu	*!		*
d. ➜ tu		*	

Above, we see that complex onsets in general are permitted (*Faith >> *Complex*); however, s-stop clusters are outlawed because they violate *Sonority Sequencing (SSP)*.

Barlow (2001) argues that variation among children in acquisition of these clusters is expected, given the fact that /s-stop/ clusters are not true complex onsets, but a sequence of adjunct followed by onset. Therefore, the child who possesses stop-liquid clusters, but not s-stop clusters would have **Adjunct* in addition to *SSP* ranked above *Faith*.[3] The existence of such a constraint allows also for a child who has the opposite pattern: initial s-stop clusters, but no initial stop-liquid clusters. The grammar of this child is displayed in (2).

(2)   **Complex >> Faith >> *Adjunct*

/blu/	*Complex	Faith	*Adjunct
a. ➜ bu		*	
b. blu	*!		
/stu/	*Complex	Faith	*Adjunct
c. ➜ stu			*
d. tu		*!	

The tableaux in (1) and (2) show that with **Adjunct*, both acquisition orders are possible. The fact that stop-liquids are often acquired before s-stops may then be related to the

---

[2] Another proposal was to consider *sC*-initial clusters as complex *segments* rather than complex *onsets* (Selkirk, 1982).
[3] As pointed out by Barlow (2001), we must assume the existence of a set of inviolable constraints which prevent other consonants besides /s/ from being appendices.

greater frequency of complex onsets (=obstruent-liquids/obstruent-glides) in English relative to /s/ + Adjunct sequences. The additional assumption here is that the frequency with which constraints are violated in the adult language (=the input to the child's grammar) influences the rate of constraint demotion (Boersma, 1997; Boersma & Levelt, 2000; Levelt & van de Vijver, 2004).

2.2.2. Syllable structure vs. faithfulness constraints in medial position

Intervocalic consonant clusters differ from word initial clusters in that they can either be syllabified as a *complex onset* or as a *coda + onset* sequence. In the first case, **Complex* is violated; in the second, at least, *NoCoda*. Since both codas and complex onsets are legal in English, *Faith* must be ranked above both *NoCoda* and **Complex* in the adult grammar. A further ranking between *NoCoda* and **Complex* would seem to determine the syllabification of intervocalic /CC/ sequences, as exemplified below:

(3)   a. Faith >> *Complex >> NoCoda   [VC.CV]   e.g. *pas.ta*
        b. Faith >> NoCoda >> *Complex   [V.CCV]   e.g. *pa.sta*

The existence of the constraints *NoCoda* and **Complex* predicts that some languages maximize intervocalic onsets (3b), while others minimize them (3a) (see Kirk 2001, Treiman & Zukowsky 1990, for example, for discussion of the *Onset Maximization Principle*). The implication of these constraints for child speech is a stage of development in which a legal word-initial cluster /C_1C_2/ is found word medially, but not word initially. This is shown in (4).

(4)   **Complex >> Faith >> NoCoda*

/CCV/	*Complex	Faith	NoCoda
a. ➜ CV		*	
b. CCV	*!		
/VCCV/	*Complex	Faith	NoCoda
c. ➜ VC.CV			*
d. V.CCV	*!		
e. V.CV		*!	

2.2.3. Sonority and syllabification decisions

While the constraint rankings in (3) determine whether onset maximization takes place in general in the language, they are not sufficient to account for cases in which medial cluster syllabification is non-uniform across cluster type.   For example, in Italian, intervocalic obstruent-liquid clusters are tautosyllabic, while s-initial clusters are generally considered to be heterosyllabic, e.g. *a.pro 'I open' vs. pas.ta 'pasta'*. Morelli (2003) uses the ranking **Complex >> NoCoda* to capture the default heterosyllabicity pattern; in intervocalic obstruent-liquid clusters, on the other hand, onsets are maximized because the alternative syllabification (e.g. Vp.lV) contains a marked syllable contour, by the *Syllable Contact Law*.

(5)   *Syllable-Contact*: Sonority falls across a syllable boundary (Gouskova, 2000; Hooper, 1976; Kirk, 2001; Murray & Vennemann, 1983).

The implication of the following constraint (or family of constraints – see Gouskova, 2000) is that in a stage in which word initial clusters are disallowed (by **Complex*), intervocalic clusters consisting of /s/ followed by a stop are preferred to those consisting of a stop followed by liquid.

(6)   **Complex >> Syllable-Contact >> Max*

/VspV/	*Complex	Syllable-Cont	Max
a. V.stV	*!		
b. ➔ Vs.tV			
c. V.pV			*!
/VprV/	*Complex	Syllable-Cont	Max
d. V.prV	*!		
e. Vp.rV		*!	
f. ➔ V.pV			*

The tableau in (6) reflects a grammar in which complex onsets are altogether excluded in word-initial position under the influence of **Complex*. Medial clusters are allowed as long as they are heterosyllabic. And in this case, s-stop clusters are produced, while stop-liquid clusters reduced, on account of *Syllable-Contact*.[4,5]

2.2.4. Effect of prosody

Stressed syllables are obligatorily heavy in some languages. One way to satisfy this requirement is through the lengthening of vowels or of consonants (see Kristoffersen, 1999 for Swedish; Nespor, 1993 for Italian, etc.). However, as long as consonants contribute weight, an intervocalic sequence of two adjacent consonants following a stressed syllable may satisfy *Stress-to-Weight* without vowel lengthening if the consonants are parsed in separate syllables.

(7)   *Stress-to-Weight*: Stressed syllables are minimally bimoraic

Findings from syllabification judgment tasks for adult English reveal a greater tendency to place the syllable boundary between the consonants in a /CC/ cluster when the preceding syllable is stressed and lax (=light) than when it is unstressed or stressed

---

[4] Following Gouskova (2000), *Syllable-Contact* is not a unitary constraint, but a fixed ranking of constraints, whereby the highest ranking constraint disallows the maximum sonority increase across a syllable boundary. While heterosyllabic s-stop clusters do better than heterosyllabic stop-liquid clusters on *Syllable-Contact*, clusters of the former type do worse than other clusters which present a fall greater than 1 on the sonority scale. In this paper, I consider stop-liquid clusters to violate *Syllable-Contact*, while s-stop clusters violate *Syll-Con$_{[>1]}$*, requiring the fall in sonority to be more than one.

[5] *NoCoda* must be ranked below *Max*; this is not a problematic assumption, given the fact that codas are relatively early to emerge in the speech of children acquiring English.

Della Chambless

and tense (=heavy) (e.g. Treiman & Danis, 1988; Treiman & Zukowski, 1990).[6] The relative ranking between *Stress-to-Weight* (*S-to-W*) and other markedness (e.g. **Geminate, *VV*) and faithfulness (e.g. *Dep-Mora*) constraints determines whether and how this constraint is satisfied in a language.

With regard to child language, *Stress-to-Weight* has a potential role in producing asymmetries in the production of intervocalic consonant clusters. Even if the ranking of *Markedness* >> *Faithfulness* prevents a cluster from occurring intervocalically (where *Markedness* includes constraints such as **Complex*, members of the *Syll-Con* family, *NoCoda*, etc.), the higher ranking of the two markedness constraints *Stress-to-Weight* and *Onset* result in both coda and onset position being filled, in the case in which two or more intervocalic consonants following a lax stressed vowel are present in the input. An example of such a grammar is given in (8).

(8)   *Stress-to-Weight, Onset >> Syll-Con[> 1] >> Max*

/V[lax]stV/	Stress-to-Weight	Onset	Syll-Con[> 1]	Max
a. ➔ V[lax]s.tV			*!	
b. V[lax].tV	*!			*
c. V[lax]C.V		*!		*
/V[tense]stV/	Stress-to-Weight	Onset	Syll-Con[> 1]	Max
d. ➔ V[tense].CV				*
e. V[tense]s.tV			*!	

The ranking in tableau (8) shows a grammar in which syllable contact restrictions prevent a heterosyllabic cluster [st] from emerging in (d), because the decrease in sonority is too small between them. However, in a word in which the cluster is preceded by a lax (stressed) vowel (8a), the cluster appears under pressure from *Stress-to-Weight*. A high ranking of *Onset* prevents satisfaction of *Stress-to-Weight* by simply syllabifying a single intervocalic consonant as a coda, rather than onset, as in (8c).[7]

2.3.   Experiment

2.3.1. Introduction

The experiment was designed to test the activity of the markedness constraints described in the previous section. We expect contextual asymmetries in acquisition order (quantified as rate of faithful cluster production) to pattern in a way consistent with interactions among those constraints.

---

[6] There is some evidence from metalinguistic tasks that a stressed syllable attracts a medial consonant even when tense (Treiman & Zukowski, 1990, Kirk, 2001). *Stress-to-Weight* alone (as currently defined) would not get us this effect, because tense vowels are considered bimoraic. Kirk (2001) accounts for this finding with the *Stress Attracts Principle*, which holds that consonants are attracted to stressed syllables.

[7] Another possible candidate which would satisfy *Onset*, *Stress-to-Weight*, and *Syllable Contact* is an output form in which the intervocalic singleton consonant is ambisyllabic.

2.3.2. Methods

Subjects

The subjects were 32 children ranging in ages from 18 to 38 months. More than half were tested in a daycare setting, while others were interviewed in their homes with parents present. All of the subjects were native English speakers.

Stimuli

Conditions manipulated were cluster type (stop-liquid or s-stop) and word environment (initial, post-lax (stressed) V, post-tense (stressed) V, or post-unstressed V). The stimuli were invented disyllabic proper names (e.g. *Stoller, Leedrow*). The *stop-liquid* clusters were /bl, br, pr, dr, kl/. The *s-stop* clusters were /sk, st, sp/. The vowels in the syllable preceding the cluster were [\, ↔, i, e and ∴]. Second syllables ended in either /C, o, i, or ∴/. There were a total of 8 stimuli for each condition.

WHAT SHOULD THE VOWELS BE?

Procedure

Children were tested on 1-4 occasions. The criterion for when to cease data collection for a child was when a minimum of 6 tokens had been gathered for each condition (e.g. br-initial, where "br" represents *stop-liquid* cluster type). The duration anticipated for each session was approximately 20 minutes. However, if the child lost interest before the end of the activity, he/she was interviewed again subsequent to that date and all data were pooled. The materials used were 20-30 stuffed animals and picture cards. The testing procedure took one of two forms. These will be described in turn.

In the first activity, animals were taken out of a bag one by one, and introduced by the experimenter. The child was instructed to say "*hello X,*" in which X stood for the name of the animal, as dictated by the experimenter. After all the animals had been taken out of the bag, the child was told to put them back in again, this time saying "*bye-bye X,*" where X was once again a stimulus name assigned arbitrarily to the animal by the experimenter. Each stimulus name was handwritten on an index card. Initially, cards were shuffled so as to assure a random ordering of names.

In the second activity the child was presented with pictures of people. The instructions were to listen to the name uttered by the experimenter for each person illustrated, repeat the name, and then put the picture card in a box. The first procedure was introduced first for all children. The second was taken up if the child was not tired after the first or if the first activity didn't engage the child's attention sufficiently.

2.3.3. Hypotheses

The hypotheses below follow from the markedness constraint discussed in the preceding section. These constraints should result in asymmetries, with some clusters preceding others in development.

(9)   *Predicted asymmetries among children still in process of acquiring clusters*

(i)   *Word position*: Some children will produce both members of cluster more frequently word medially than word initially.

(ii)   *Cluster type*: In children who do not produce any word-initial clusters, [sp] will be preferred to [br] *intervocalically*.

(iii)   *Interaction cluster type with position*: /sp/ will do better relative to /br/ in medial position than in initial position.

(iv)   *Prosody*: Clusters following lax stressed vowels will be produced faithfully more often than the same clusters following tense stressed vowels or unstressed vowels.

Prediction (i) is possible on account of syllable structure constraints such as *NoCoda* and **Complex*. If **Complex* outranks *Faith*, clusters will not be allowed word initially. If *Faith* outranks *NoCoda*, a cluster is possible word medially as long as it is heterosyllabic. This should hold even if the cluster is permitted word initially (and thus held to be a complex onset) in the adult language.

Prediction (ii) is based upon the family of *Syllable-Contact* constraints. A heterosyllabic [sp] cluster fares better on *Syllable Contact* than its [bl] counterpart (we'll assume that [sp] violates *Syll-Con[>1]*; while [bl] violates the same constraint, in addition to other higher rankings ones: *Syll-Con$_{[>-2]}$*, *Syll-Con$_{[>-1]}$*, *Syll-Con$_{[>0]}$*, etc. (Gouskova, 2000)). Therefore, assuming there is a stage in which the only intervocalic clusters allowed are heterosyllabic ones (**Complex >> Faith >> NoCoda*), we predict that [sp] clusters are subject to less deletion than [bl] ones.

Prediction (iii) follows from *Sonority Sequencing (SS)* and **Adjunct*. Violation of both of these constraints can be avoided in intervocalic position, because the /s/ of the s-stop cluster can be syllabified as a coda. Therefore, the advantage which stop-liquid clusters might have over s-stop clusters in initial position should be lessened in word medial position.

Prediction (iv) is a result of *Stress-to-Weight*. This constraint can be satisfied by a bimoraic (long) vowel or the presence of a coda consonant. Therefore, we expect clusters to be produced more consistently in the post-lax environment than in the post-tense environment, on the assumption that a candidate with a heterosyllabic cluster can satisfy *Stress-to-Weight* independent of the ranking of *Faith* with respect to other markedness constraints.[8]

---

[8] A further constraint prefers a stressed CVC with a lax V to the same syllable with a tense V: **TrimoraicSyllable* (Kager, 1999).

2.3.4. Results

The experimental sessions were recorded using a Sony MD-X-70 recorder and were transcribed at least twice for accuracy. Cluster responses were coded in the following way: realized faithfully or deleted. Responses which could not be transcribed with confidence after 3 attempts (mostly due to inaudibility of child's utterances) were discarded. The following criteria were applied to eliminate subjects from the group data analyzed:

- Subjects regularly exposed to a 2nd language (=4)
- Hearing disorder (=1)
- Insufficient data gathered due to non-cooperation /sickness, etc. (=2)
- Clusters produced too consistently (=7)
- Clusters produced too inconsistently (=5)

A subject who deleted both cluster types less than 20% of the time in all environments was considered too advanced for participation. A child was considered not advanced enough for inclusion if deletion occurred more than 80% of the time in both cluster types across all contexts. These subjects were not included in the analysis because their speech couldn't provide clues as to asymmetries in cluster production: clusters were either all present everywhere or not present anywhere. Fourteen children remained once subjects were eliminated for the reasons given above.

For the purposes of this analysis, two categories were conflated: fully faithful realization (e.g. /br/ = [br]) and almost faithful realization (eg. /br/ = [bw]). Gliding was common in many children's speech, and furthermore, often targeted simple liquid onsets as well. For each subject, percentages of utterances in which faithful realization of the cluster occurred were calculated according to cluster type (*stop-liquid* vs. *s-stop*) and context (*post-lax*, *post-tense*, *post-unstressed*, and *word-initial* position). Percentages were calculated as a ratio of number of utterances of a certain type in which both consonants in the cluster were produced divided by the total number of times in which a word containing a sequence of a particular type was attempted.

The *unstressed* category was ultimately excluded from the analysis. Initial unstressed syllable deletion occurred in the speech of more than two thirds of the subjects analyzed. Thus, the disyllabic utterances with final stress (e.g. Sabrine) were often produced as monosyllabic (e.g. [bin] or [brin]). To have included these productions in the *unstressed* category would not have been truly representative of that condition, in so far as they resemble the *word-initial* context rather than a *word-medial* one. While one possibility was to remove all truncated responses from the unstressed category and to code them as instances of the initial category, the results of this classification were problematic as well. While there was no great change in the initial condition once the truncated exemplars were added, there were undesirable results for the *unstressed* condition. For some children, once all truncated utterances were removed from the *unstressed* category, the number of exemplars left in that category were too few to be reliably representative. For these reasons, I eliminated altogether the *unstressed* category from this analysis.

Della Chambless

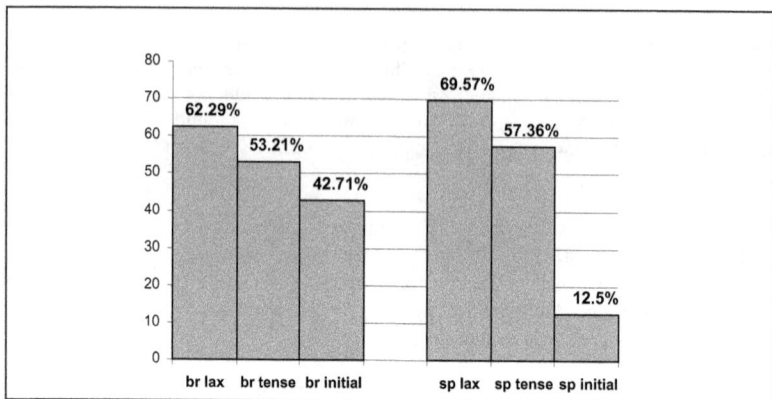

Figure 1. Percentage of faithful production of clusters across position, by type.

Figure 1 shows the positional asymmetries for both /br/ and /sp/ clusters: the most
frequent faithful realization occurs following the lax (stressed) vowel and least frequent
occurs in initial position. A 2 x 3 (cluster type x prosodic context) within-subjects
repeated measures analysis of variance (ANOVA) was carried out. The main effect of
cluster type was not significant, $F(1,13) = 0.640$, while the main effect for prosody was
highly significant, $F(2,12)= 24.88$, $p < .001$. There was also a significant interaction of
cluster type and prosody, $F(2,12) = 8.489$, $p<.001$.

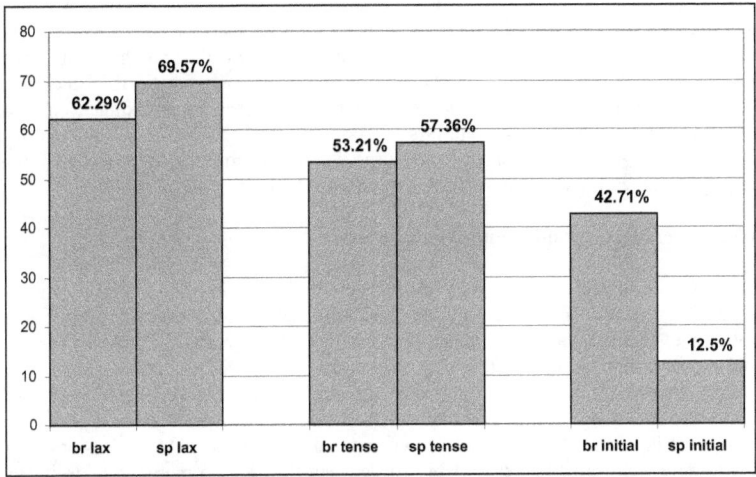

Figure 2. Percentage of full /CC/ realization by prosody (1=lax, 2=tense, 3=initial) and cluster (L=br; R=sp) N=14.

Figure 2 shows the cluster type asymmetry by position. Whereas /sp/ is favored over /br/ in both medial conditions, the opposite is the case for initial position. Only the latter finding (br > sp word initially) is significant, however. There is no overall main effect of cluster type, but an interaction between cluster type and prosody, precisely because the greater tendency to reduce one cluster with respect to another was reversed when the cluster's position in the word changed.

2.3.5. Discussion

Results from the analysis lead to the following observations and subsequent conclusions regarding the roles of the constraints under question.

2.3.5.1. Word initial position: Sonority sequencing and *Adjunct

The results show a striking asymmetry of cluster type in <u>word initial</u> position. Between [sp] and [br], the former is produced much less reliably. This is not clearly predicted by the constraint set since both orders have been shown to be possible on the basis of *Adjunct*, among others (see Barlow, 2001 for an OT analysis).[9] One explanation for the asymmetry may be found in the frequency of clusters of the obstruent-approximant class

---

[9] Fikkert (1994) discusses such variability in order of acquisition as resulting from a difference in parameter setting, with children acquiring s-stop clusters first allowing for adjuncts but no clusters, while children with obstruent-approximant clusters and no s-stop clusters have the cluster parameter ON and Adjunct parameter OFF.

with respect to members of the /sC/ class. Following Boersma (1997), Boersma & Levelt (2000), Levelt & van de vijver (2004), etc., I assume that the speed with which constraints are demoted depends on the rate at which they are violated in the ambient (i.e. adult) language. A search performed of child-directed speech on a few CHILDES corpora reveals a greater frequency of *Complex* violators (i.e. Obstruent-approximant clusters) than *Adjunct* violators (i.e /sC/ clusters).[10] Assuming that *Complex* is demoted at a faster rate than *Adjunct*, perhaps it is not surprising that word initial stop-liquid clusters emerge prior to word initial s-stop clusters more often than not.

2.3.5.2. Intervocalic position and syllable contact

In the intervocalic context, there is not a significant difference in frequency of deletion in [sp] and [br] clusters. If [br] and [sp] were both only syllabified as onsets word medially, we would expect the same asymmetry found in 2.3.5.1. – greater faithfulness of stop-liquid clusters than s-stop clusters. On the contrary, /sp/ clusters are produced more faithfully than /br/ clusters in both medial conditions (see Figure 2), although such differences are not significant. The implication is that the disappearance of a preference for [br] over [sp] in medial position, when compared to initial position, is a result of [sp] being a more well-formed heterosyllabic cluster than [br]. Furthermore, of subjects who did not produce word initial clusters at all (on the basis of deleting more than 80% of the clusters), three out of four produced word medial s-stop clusters at a higher rate than stop-liquid clusters. I suggest that these findings are evidence for *Syllable-Contact*.

2.3.5.3. Vowel quality: Stress-to-Weight

Following a stressed lax vowel, cluster reduction occurs less often than following a tense one: *Stress-to-Weight* can accommodate this result. Since the constraint on bimoracity is satisfied already by a tense vowel, it is in the post-lax condition that in addition to *Faith*, markedness constraints (i.e. *Onset* and *Stress-to-Weight*) apply pressure to preserve both members of the cluster.

To sum up, the results of this experiment attest to the role of sonority and syllable weight in predicting *order of acquisition* of particular clusters, with cluster type and word environment as dependent variables. The next section examines ways in which similar constraints produce positional asymmetries in *cluster reduction* choices.

3.   Cluster reduction: Positional asymmetries

3.1.   Introduction

In the speech of children who reduce consonant clusters, the choice of which consonant (e.g. C1 or C2) to preserve is not random. Sonority is one factor which explains word-initial patterns (e.g. Baertsch, 2002; Gnanadesikan, 2004; Ohala, 1996; Pater & Barlow,

---

[10] Counts were drawn from the following CHILDES corpora: *MacWhinney* (MacWhinney, 2000), *Bates* (Bates et al, 1988; Carlson-Luden, 1979), and *Bloom* (Bloom, 1970). The token frequency for obstruent-approximant clusters vs. /sC/ clusters was found to be 1380 vs. 908.

2003). Intervocalic clusters are expectedly subject to similar constraints, although predictions are less straightforward, given the ambiguity of the syllabic affiliation of an intervocalic consonant. This investigation uses data from the previous experiment to address predictions outlined in the following section.

3.2.   Background and predictions

3.2.1.   Word edge cluster reduction

Word initially, both *stop-liquid* and *s-stop* clusters reduce to the stop. The frequent explanation for such a preference is the relative sonority of the members of the cluster. In onset position, low sonority consonants are preferred, while in coda position, high sonority consonants are favored (Clements, 1990). These preferences are represented by the following fixed rankings of constraints (Ohala, 1996; Pater, 1997; Prince & Smolensky, 1993):

(10)   **Glide-ONS >> *Liquid-ONS >> *Nasal-ONS >> *Fricative-ONS*
(11)   **Stop-CODA >> *Fricative-CODA >> *Nasal-CODA >> *Liquid-CODA*

Experimental evidence for such rankings in acquisition comes from, among others, Ohala (1996). In utterances in which one of a two member cluster was deleted, the stop was found to be preserved at a greater rate than the /s/ word initially, while the clusters reduced to the /s/ more often word finally. This finding is consistent with the rankings in (10) and (11).

3.2.2.   Word medial cluster reduction

Given the constraint rankings in (10) and (11), predictions as to which consonant will be retained when medial clusters are reduced hinge upon assumptions as to the syllabic role of the intervocalic consonant. If it is an onset, we expect the least sonorous consonant of the cluster to be preserved; if it is a coda, we expect the more sonorous consonant to be preserved.

The status of intervocalic consonants in English is subject to some debate. Although the existence of constraints *NoCoda* and *Onset* converge upon the syllabification /VCV/ → [V.CV], it has often been claimed that an intervocalic consonant preceding an unstressed syllable is a coda or is ambisyllabic (Giegrich, 1992; Gussenhoven, 1986; Hammond, 1999; Pulgram, 1976; Selkirk, 1982).

If the intervocalic consonant is a coda, then we predict by (11) that when intervocalic s-stop clusters are reduced, the consonant preserved will be an /s/, rather than a stop. If, on the other hand, the consonant is simultaneously both a coda and an onset, then whether the fricative or stop gets selected will be a function of the interaction between the rankings between (10) and (11). If, for example, **Stop-CODA* outranks **Fricative-ONS*, then the /s/ will be selected over the stop. This grammar is illustrated below.

Della Chambless

(12)   Possible grammar: *Syll-Cont[> 1] >> *Stop-CODA >> *Fric-ONS*[11]

/VspV/	Syll-Contact$_{[> 1]}$	*Stop-CODA	*Fricative-ONS
☞a. σ σ \\/ V s V			*
b. σ σ \\/ V p V		*!	
c. σ σ \ / Vs.pV	*!		

Following the hypothetical grammar in (12), if we accept the idea that intervocalic singleton consonants are sometimes codas, then we predict that the /s/ will be selected over the /stop/ more often word medially than word initially, because in the former environment, it is not unambiguously an onset.

Syllabification tasks (Treiman & Zukowsky, 1990; Treiman & Danis, 1988; Zamuner & Ohala, 1998) have shown that the propensity of an intervocalic consonant to be syllabified as the coda of a preceding syllable depends on the prosody of the two syllables flanking the consonant. An intervocalic consonant is more likely to be attracted to the preceding syllable as opposed to a following syllable if the preceding syllable is stressed and contains a lax vowel than if it is stressed with a tense vowel or is unstressed. This results from the principle of bimoracity, reflected in the constraint *Stress-to-Weight* (Prince & Smolensky, 1993).

The way that this constraint interacts with sonority constraints to produce asymmetries in reduction patterns is as follows: given an intervocalic cluster containing C1 and C2, which differ in relative sonority, the more sonorous of the two consonants is predicted to be preserved more often following a stressed lax syllable than in other conditions, given the greater likelihood in such a condition of the consonant acting as a coda.

3.3.   Experiment (Cluster reduction data analysis)

3.3.1.   Introduction

The subjects, stimuli and procedure of the experiment were outlined in section 2.3.2. The reduction analysis performed and reported on in this section, dealt not with percentage of clusters produced correctly, but rather with the rates at which C1 or C2 was preserved in cases in which reduction occurred.

Hypotheses regarding cluster reduction tendencies are repeated below:

---

[11] A necessary assumption is that any *Ambisyllabicity* constraint is low ranked.

(i) Given the possibility for an intervocalic consonant to be syllabified as a coda intervocalically, it was expected that the more sonorous consonant (e.g. /s/ from s-stop clusters) would be selected over the stop more often word medially than word initially (*Stop-CODA* >> *Fricative-CODA*).

(ii) The existence of *Stress-to-Weight* predicts that the intervocalic consonant will be a coda more often following a stressed syllable containing a lax vowel than a stressed syllable containing a tense vowel, or an unstressed syllable. Therefore, given *Stop-CODA* >> *Fricative-CODA*, the more sonorous consonant (s) is predicted to be preserved more often in the post-lax condition than in the post-tense and post-unstressed condition.

3.3.2. Results – Cluster reduction

/s/-stop clusters

To be included in this analysis subjects had to have at least 4 exemplars of reduced clusters for each group – post-lax, post-tense, initial, and unstressed. Ten subjects met this criterion. The percentages reported were calculated as percentage out of all reduced clusters in which a particular consonant (**s** or **stop**) was preserved.

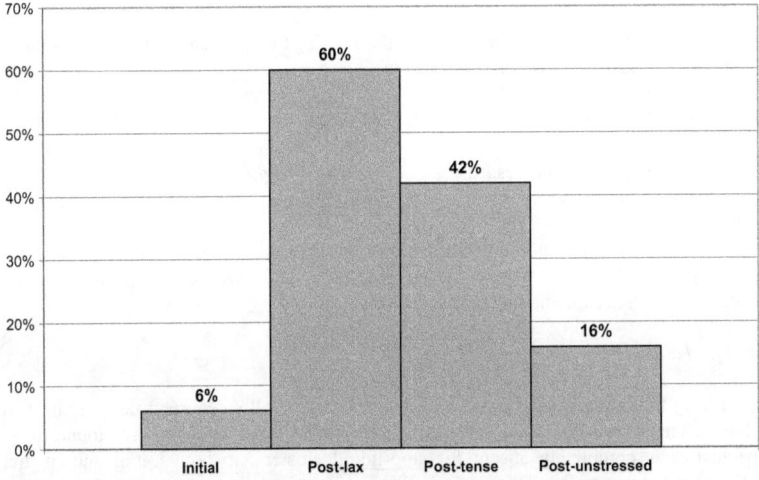

Figure 3. Percentage of all reduced clusters in which /s/ is preserved.

A repeated measures ANOVA revealed a main effect of position, $F_{(3, 7)} = 9$, $p < .001$. Furthermore, the following pairwise comparisons were significant:

(13)   Significant differences among conditions:

   a. Post-lax > Initial   F $(1,9) = 22.8$, p $< .001$
   b. Post-lax > Post-tense   F $(1,9) = 9.8$, p $< .012$
   c. Post-lax > Post-unstressed F $(1,9) = 14.4$, p $< .004$
   d. Post-tense > Initial   F $(1,9) = 6.7$, p $< .029$

The fact that /s/ surfaces more often medially than initially would seem to suggest that the intervocalic /C/ is sometimes a coda. The fixed ranking from (11) *(*Stop-CODA >> *Fricative-CODA)* accounts for the choice of /s/ over the stop. The significant difference in 13(b) suggests that an intervocalic /C/ is more likely to be a coda (or ambisyllabic) when the previous syllable is stressed and light. This amounts to the activity of *Stress-to-Weight* and **Stop-CODA*. However, the fact that /s/ is retained more often in the post-tense condition than in the initial condition (13d) is not explainable by *Stress-to-Weight* alone. Something else is needed.

One possibility is that this result reflects a difference between initial and medial position which is independent of syllable structure. The intervocalic environment is one in which lenition typically occurs. The constraint in (14) captures the fact that continuants are favored over stops intervocalically (Kirchner, 1998).

(14) $*VC_{[stop]}V$ Only continuants can occur in the intervocalic environment.

$*VC_{[stop]}V$ along with *Stress-to-Weight* and the rankings in (10) and (11) are capable of explaining the results in (13), repeated below:

(13)   a. Post-lax > Initial $*VC_{[stop]}V$
        b. Post-lax > Post-tense   *Stress-to-Weight*
        c. Post-lax > Post-unstressed   *Stress-to-Weight*
        d. Post-tense > Initial $*VC_{[stop]}V$

The activity of $*VC_{[stop]}V$ should result in a split between initial and all medial conditions. We still get a difference in post-lax vs. post-tense and post-lax vs. post-unstressed conditions, which is explainable by *Stress-to-Weight*.

Stop-liquid clusters

It is striking that intervocalic *stop-liquid* clusters (across all categories) never reduce to the liquid. This means that the initial-medial asymmetry in reduction patterns found for s-stop clusters is completely absent for stop-liquid clusters: in both initial and medial position /br/ reduces to [b]. This outcome is surprising in the face of the fixed constraint ranking in (11), which shows that a liquid is a better coda than a fricative. It is also inconsistent with results from some syllabification studies which show that sonorants are more likely to affiliate to a previous syllable than obstruents (e.g. Treiman & Danis, 1988).

A possible explanation for such a finding is that the intervocalic /C/ is an onset even when it's also a coda (i.e. ambisyllabic), subject, therefore, to a high ranked constraint against liquid onsets. If **Liq-ONS* outranks **Stop-CODA*, a stop-liquid cluster will be reduced to the stop, assuming that this single intervocalic consonant is produced either as an onset or onset AND coda, but never as a coda by itself.

(15)   *Stress-to-Weight, Onset, *Liq-ONS >> *Stop-CODA*

/V$_{[lax]}$blV/	Stress-to-Weight	Onset	*Liq-ONS	*Stop-CODA
a. ➔σ σ ∨ V b V				*
b. σ σ ∨ V l V			*!	
c. σ σ ∧ | V l V		*!		
d. σ σ | ∧ V b V	*!			

While the ranking in (15) accounts for grammars in which the liquid is not selected over the stop word medially, the opposite grammar should emerge when the ranking between **Liq-ONS* and **Stop-CODA* is reversed.

(16)   *Stress-to-Weight, Onset, *Stop-CODA >> *Liq-ONS*

/V$_{[lax]}$blV/	Stress-to-Weight	Onset	*Stop-CODA	*Liq-ONS
a. ➔σ σ ∨ V l V				*
b. σ σ ∨ V b V			*!	
c. σ σ ∧ | V l V		*!		

Since the grammar in (16) is consistent with one permutation of the constraints in (10) and (11), we should expect at least a small difference on average between medial and initial position, similar (if not smaller) to what was found for s-stop clusters. The fact that /l/ is never selected, i.e. that we find 100% retention rate of the stop in all categories, could be a result of the small number of subjects in the experiment.[12]

---

[12] There is another potential explanation for the asymmetry between stop-liquid and s-stop clusters in the choice of C1 vs. C2 in medial position. If inputs are prosodified at least at this stage of learning, C1 may have an edge over C2, when it is a coda to a preceding stressed syllable. The constraint at work would be something like *Head Faithfulness* (see Goad & Rose, 2004), requiring elements of a head syllable in the input to be preserved in the output. Such a constraint would gives /s/ an extra edge over the stop in s-stop

4.   Conclusion

The clusters that I examined in this paper were a subset of the possible word-initial clusters in English: *stop-liquid* and *s-stop*. Such clusters are also permitted word medially, and, in fact, important positional differences in cluster production and reduction were predicted and demonstrated to exist. I argued that initial vs. medial differences and differences within the medial (prosodic) conditions were a product of the greater syllabification possibilities afforded of intervocalic consonants – singletons and clusters alike. Furthermore, I suggest that constraints on stress and syllable weight interact with sonority constraints to produce such results.

References

Alderete, J. (1995). Faithfulness to prosodic heads. Ms. University of Massachusetts at Amherst.

Barlow, J. (1997). *A Constraint-based Account of Syllable Onsets: Evidence from Developing Systems*. PhD dissertation. Indiana University.

Barlow, J. (2001). The structure of /s/-sequences: evidence from a disordered system. *Journal of Child Language* 28, 291-324.

Barlow, J. & J. Gierut (1999). OT in phonological acquisition. *Journal of Speech, Language and Hearing Research.* 42. 1482-1498.

Barlow. J. (2003). Variation in cluster production patterns by Spanish-speaking children. Handout from the *28th Annual Conference of Language Development,* Boston, Massachusetts..

Beckman, J. (1997). *Positional Faithfulness*. Ph.D. dissertation. University of Massachusetts at Amherst.

Bernhardt, B. and J. Stemberger (1998). *Handbook of Phonological Development from the Perspective of Constraint-based Nonlinear Phonology*. San Diego: Academic Press.

Blevins, J. (1995). The syllable in phonological theory. In the *Handbook of Phonological Theory* (ed. J. Goldsmith), Cambridge, MA: Blackwell Publishers.

Boersma, P. and C. Levelt (2000). Gradual constraint ranking learning algorithm predicts acquisition order. In the *Proceedings of the 30th Child Language Research Forum,* ed. E. Clark, p. 229-37. Stanford, CA: CSLI.

Chambless, D. (2004). Asymmetries in initial and medial cluster acquisition. *Proceedings of the 28th Annual Conference on Language Development.* Somerville, MA: Cascadilla Press.

Compton, A. & M. Streeter (1977). Child phonology: data collection and preliminary analyses. In *Papers and Reports on Child Language Development* 7. Palo Alto, California: Stanford University.

---

clusters (assuming s-stop clusters to be heterosyllabic), but it does nothing to affect the preference for the consonant of lower sonority in stop-liquid clusters, since it is the stop which is C1 and part, potentially, of the preceding stressed syllable. Whether or not the order of the two consonants (C1 or C2) makes a difference in reduction preferences needs to be subjected to further testing. For example, reduction of intervocalic s-stop clusters must be compared to reduction of intervocalic stop-s clusters. This such a comparison is currently under investigation.

Fikkert, P. (1994). *On the Acquisition of Prosodic Structure*. PhD dissertation. Leiden University. Published 1994, The Hague: Holland Academic Graphics.

Gierut, J. (1999). Syllable onsets: clusters and adjuncts in acquisition. *Journal of Speech Language & Hearing Research*. 42. 708-26.

Gnanadesikan, A. (2004). Markedness and faithfulness constraints in child phonology. In R. Kager, J. Pater & W. Zonneveld (eds) *Fixing Priorities: Constraints in Phonological Acquisition*. Cambridge: Cambridge University Press.

Goad, H. & Y. Rose (2004). Input elaboration, head faithfulness and evidence for representation in the acquisition of left-edge clusters in West-Germanic. In R. Kager, J. Pater & W. Zonneveld (eds) *Fixing Priorities: Constraints in Phonological Acquisition*. Cambridge: Cambridge University Press.

Gouskova, M. (2000) Falling sonority onsets, loanwords and syllable contact. *Ms.* University of Massachusetts at Amherst.

Kirk, C. (2001). *Phonological constraints on the segmentation of continuous speech*. Ph.D. dissertation. University of Massachusetts at Amherst.

Levelt, C. and Van de Vijver (2004). Syllable types in cross-linguistic developmental grammars. In *Fixing Priorities: Constraints in Phonological Acquisition*, ed. R. Kager, J. Pater & W. Zonneveld. Cambridge: Cambridge University Press.

Morelli, F. (2003). The relative harmony of /s+stop/ onsets: obstruent clusters and the sonority sequencing principle. In *The Syllable in Optimality Theory*, ed. C. Féry & R. van de Vijver.

Nordstrom, B. (2001). A study in child consonant harmony. Undergraduate honor's thesis. University of Massachusetts, Amherst.

Ohala, D. (1996). *Cluster Reduction and Constraints in Acquisition*. Ph.D. dissertation, University of Arizona.

Ohala, D. (2004). Word Boundary Effects in the Reduction of Consonant Sequences in Young Children's Speech. Handout from Child Phonology Conference: May 14, 2004.

Prince, A. & P. Smolensky (1993). Optimality theory: Constraint interaction in generative grammar: Technical Report no. 2. New Brunswick, NJ: Rutgers Center for Cognitive Science, Rutgers University.

Selkirk, E. (1982). The syllable. In H. van der Hulst & N. Smith (Eds.), *The Structure of Phonological Representations* (pp. 337-383). Dordrecht, Netherlands: Foris.

Department of Linguistics
South College
University of Massachusetts
Amherst, MA 01003

dchamble@linguist.umass.edu

Lateral acoustics and phonotactics in Australian languages[*]

Kathryn Flack

University of Massachusetts, Amherst

1.   Lateral distribution and possible analyses

Australian languages have a large number of contrastive coronal places of articulation; three or even four coronal stops and nasals are common. Australian languages often have laterals articulated at each of these coronal places, and as such have more contrastive voiced lateral approximants than any other language group (Ladefoged and Maddieson 1996: 185). Panyjima's four laterals represent the maximal lateral inventory; other Australian languages have all or a subset of these laterals.

(1)   Panyjima phoneme inventory (Dench 1991)

	Labial	Dental	Apico-alveolar	Retroflex	Palatal	Velar
Stop	p	t̪	t	ʈ	c	k
Nasal	m	n̪	n	ɳ	ɲ	ŋ
Lateral		l̪	l	ɭ	ʎ	
Rhotic			ɾ	ɻ		
Glide	w				j	

Laterals are often subject to phonotactic restrictions which prevent them from surfacing word-initially or postconsonantally in many Australian languages, as in (2).

---

[*] Thanks to John Kingston, John McCarthy, Michael Becker, Shigeto Kawahara, Ehren Reilly, the participants in UMass Second Year Seminar and Phonology Group, and the audience at NELS 35 for suggestions and helpful discussion; also to Tim Beechey, Claire Bowern, Gavan Breen, and Rob Pensalfini for language data and advice.

(2)   a. <u>Panyjima</u>   (Dench 1991)

           Laterals: ɭ l ̪l ʎ   Word-initial: *ɭ *l *l̪ *ʎ   Postconsonantal: *ɭ *l *l̪ *ʎ

       b. <u>Anindilyakwa</u>   (Leeding 1989)

           Laterals: ɭ l ʎ   Word-initial: ɭ l ʎ   Postconsonantal: *ɭ *l *ʎ

The appearance of laterals across these contexts is implicational: if a language allows laterals to appear postconsonantally, it allows them word-initially; if a language allows laterals word-initially, it allows them postvocalically.

In current phonological theory (specifically, in Optimality Theory; Prince and Smolensky 1993), two basic approaches are commonly used for explaining phonotactic restrictions. A formal, sonority-based approach (Gouskova 2002, 2003; Smith 2002) makes crucial use of restrictions on sonority in particular segmental or prosodic contexts. A fundamentally acoustically-based framework like Licensing by Cue (Steriade 1997, 1999), on the other hand, allows the interaction of segmental acoustics and contextual acoustics to determine the environments in which particular segments appear.

These two approaches are not entirely opposed in their premises. Many "formal" constraints are grounded in articulation and/or perception, and Licensing by Cue grammaticizes perceptual scales in constraints which interact with the rest of a phonological grammar. A key difference is the directness with which the grammar refers to the physical and perceptual properties of language. Licensing by Cue claims that speakers have direct access to the relative perceptability of any segment in any context; this information is encoded in the P-map (Steriade 2001). This is used to generate hierarchies of constraints which preferentially license segments in perceptable contexts.

Formal theories of contextual licensing, e.g. positional augmentation (Smith 2002) and positional faithfulness (Beckman 1998), are also grounded in functional pressures, but these pressures act less directly than they do in Licensing by Cue. Constraints in these, more formally-oriented frameworks are simply given as primitives in a grammar, rather than being built based on knowledge of acoustic and perceptual salience. The constraints can license a particular segment (or class of segments) e.g. word-initially because that segment is easy to perceive in that context, or for additional reasons: such licensing may facilitate word boundary identification, or lexical recognition. A major goal of current work in phonology is to understand the ways in which these formal and functional mechanisms interact, and to determine the range of processes for which each type of analysis is truly explanatory.

Australian laterals are an important test of the explanatory scope of these types of analyses, as both acoustic and formal approaches appear to be likely explanations of their phonotactics. This is because the patterns of lateral phonotactics look like other phonotactic patterns that have been previously analyzed in each framework, as follows.

Lateral phonotactics bear a strong resemblance to those of other classes of segments whose distributions have been explained in terms of Licensing by Cue, e.g. retroflex segments in Australian and other languages. In a striking parallel to the lateral pattern described above, Steriade (1999) shows that if a language allows a retroflex contrast to surface postconsonantally, the contrast also surfaces word-initially, and that all languages with word-initial retroflexion allow it postvocalically as well. This is due to the fact that retroflexion is a left-anchored contrast, meaning that the primary acoustic distinction between apico-alveolar and retroflex consonants lies in the left-edge transitions from vowels preceding these segments. These cues are thus most robust when there are, in fact, vowels preceding the consonants; word-initial position facilitates identification of this contrast less well than does postvocalic position, but still better than does postconsonantal position (presumably because the preceding consonant masks cues which are available word-initially). In the Licensing by Cue framework, this hierarchy of left-anchored environments is expressed as $V\_ > \#\_ > C\_$. The phonotactic distribution of laterals thus strongly suggests that they might also be left-anchored, and that their asymmetrical acoustics could thus explain their phonotactics.

Formal, sonority-based analyses have also accounted for aspects of lateral phonotactics, as well as other similar patterns. Smith (2002) claims that laterals are dispreferred word-initially because, given their high sonority, they fail to attract perceptual attention to the extent that less sonorous segments do; she gives a sonority-based account of restrictions against word-initial laterals which will be described in detail in section 5. Further, the absence of high-sonority segments like laterals in postconsonantal position is a common effect of syllable-contact constraints which prefer high-sonority codas followed by (crucially) low-sonority onsets (as in Gouskova 2002, 2003, also discussed below).

The phonotactic patterns of Australian laterals, then, are interesting in that they look very similar to patterns that have been shown to follow from acoustic properties and also patterns governed by segments' sonority properties. The goal of this paper is to determine which properties, and which style of analysis, are in fact responsible for the contextual restrictions on these laterals.

To test whether lateral phonotactics may be the result of contextual Licensing by Cue restrictions, this paper will investigate whether there is evidence of acoustic left-anchoring in laterals.[1] This paper will also investigate the possibility of a formal, sonority-based analysis of these patterns. I will first describe an analysis of acoustic properties of Australian laterals which would likely show asymmetries which could support a Licensing by Cue explanation; this analysis does not reveal evidence of left-anchoring. I then sketch a formal, sonority-based account of lateral phonotactics. I conclude that despite the similarity between lateral phonotactics and those of other left-anchored segments, a formal account is a more appropriate explanation of these phonotactics than is an acoustically-based analysis. The absence of a connection between phonotactic restrictions and acoustic properties raises important questions about the predictive power of the Licensing by Cue framework.

---

[1] While there have been general objections to the Licensing by Cue framework (e.g. Gerfen 2001, Howe and Pulleyblank 2001, Kingston 2002), this paper begins by assuming that the premises of Licensing by Cue are valid and asking whether lateral phonotactics could fall into its explanatory domain; questions raised by this study about the premises of this framework are addressed in section 4.

Kathryn Flack

2.   Experimental methods

In order to test the hypothesis that Australian laterals are acoustically left-anchored, and thus that their phonotactic distribution can be explained using the Licensing by Cue framework, this study investigated acoustic properties of laterals in three languages from the central and northern Northern Territory of Australia, all of which belong to different language families. Each language has lateral phonotactics which follow the V_ > #_ > C_ hierarchy of environments.[2]

(3)   a.   <u>Ngandi</u>   (Arnhem Land; Heath 1978)
          Laterals: 1 ļ   Word-initial: 1 *ļ   Postconsonantal: 1 *ļ

       b.   <u>Jingulu</u>   (Mindi; Pensalfini 1997)
          Laterals: 1 ļ ʎ   Word-initial: 1 ļ *ʎ   Postconsonantal: *1 *ļ *ʎ

       c.   <u>Warlpiri</u>   (Pama-Nyungan; Nash 1986)
          Laterals: 1 ļ ʎ   Word-initial: 1 *ļ *ʎ   Postconsonantal: *1 *ļ *ʎ

While none of the data was recorded for the purpose of acoustic analysis, all was of high enough quality to make analysis possible. Jingulu and Ngandi data came from recordings made during linguistic fieldwork,[3] and Warlpiri data from instructional Warlpiri-language tapes (Laughren et al. 1996). Exemplars of each lateral pronounced by a single speaker of each language and flanked on each side by /a/ were selected; tokens are listed in the appendix.

The frequency and rate of formant transitions constitute the locus of asymmetry which identifies the retroflexion contrast; visual examination of lateral spectrograms, as in the representative spectrogram in figure 1, did not reveal significant asymmetries in this measure.

---

[2] The data in (3) shows that retroflex and palatal laterals are sometimes subject to independent restrictions, e.g. acoustically-based Licensing by Cue restrictions, in particular phonotactic positions. These restrictions against particular laterals are not the subject of the current investigation, which is instead concerned with patterns of licensing lateral manner. If any of a language's laterals may surface in a phonotactic position, this is evidence that lateral manner is licensed in that position. The restrictions against individual laterals are discussed in Flack (2004).
[3] The author and Rob Pensalfini recorded a Jingulu speaker in Ngukurr in 1998; Tim Beechey recorded a Ngandi speaker in Numbular in 2003.

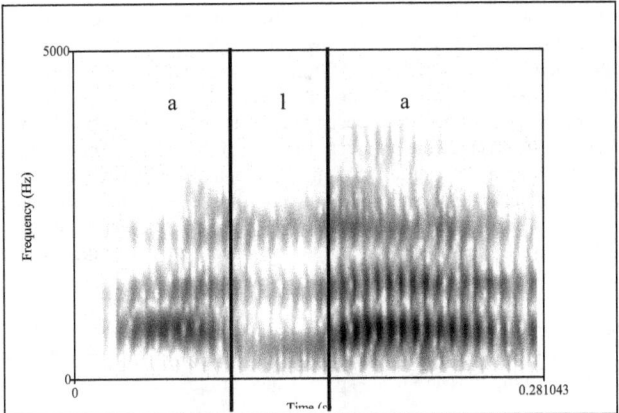

Figure 1.   Spectrogram of [ala], from Jingulu *kalara*.

Instead, this experiment proceeded following a suggestion by Stevens (1998) that there is an asymmetry in spectral amplitude at the right and left edges of laterals. Specifically, Stevens claims that there are abrupt, dramatic increases in formant amplitude during the transition from (English) laterals into following vowels, while transitions from preceding vowels into laterals show much shallower and more gradual changes in formant amplitude.

Licensing by Cue claims that such slow anticipatory changes provide listeners with more useful information about the forthcoming lateral than do abrupt changes like those at the lateral release. Therefore, if this sort of slow anticipatory decrease in intensity were consistent and significant across tokens, it would render laterals left-anchored, and thus be the basis for a Licensing by Cue account of lateral distribution.

This study measured overall spectral energy during vowel-lateral transitions and within vowels and laterals; this should reflect not only the diminished formant amplitudes in laterals but also the acoustic zeroes between formants which characterize laterals. Lateral formants are weaker than those of vowels due to the greater closure of the oral cavity. Further, the oral configuration for a lateral includes a supralingual cavity in addition to the main oral cavity; the resonances of these two cavities cancel each other out at characteristic frequencies, resulting in frequencies at which there is a notable absence of energy – an acoustic zero. Measuring the overall spectral amplitude across transitions should provide a detailed acoustic record of the rates of the articulatory changes in both oral closure and in supralingual cavity formation which precede and follow a lateral.

The edges of each lateral and flanking vowel were identified by hand, based on predictable characteristics of the waveform. After this was accomplished, all further acoustic analysis was done using Praat (Boersma and Weeninck 1992).

The positions at which spectral amplitude measurements were taken were guided by Praat's glottal-closure-identification function. This function identifies the point of maximum negative pressure – and thus the point at which the glottis is closed – within a acoustic and articulatory cycle. It achieves this by finding the pitch and thus the period of the signal at a given time, then labelling the absolute amplitude extremum within that period as a glottal closure. This procedure is recursively carried out across the signal, using the period at each closure to indicate the likely location of the next closure and thus defining the time window searched for an extremum. For a complete description of this algorithm, see the Praat manual (Boersma and Weeninck 1992). Spectral amplitude measurements were always calculated at the middle of a glottal pulse to ensure that comparable glottal states were measured. Amplitude measurements were taken at the following locations, in the manner described further below.

First, the temporal midpoint of the lateral was calculated, and the amplitude at the midpoint of the glottal pulse containing the middle of the lateral was measured. The position of this and the following measurements are indicated in figure 2.

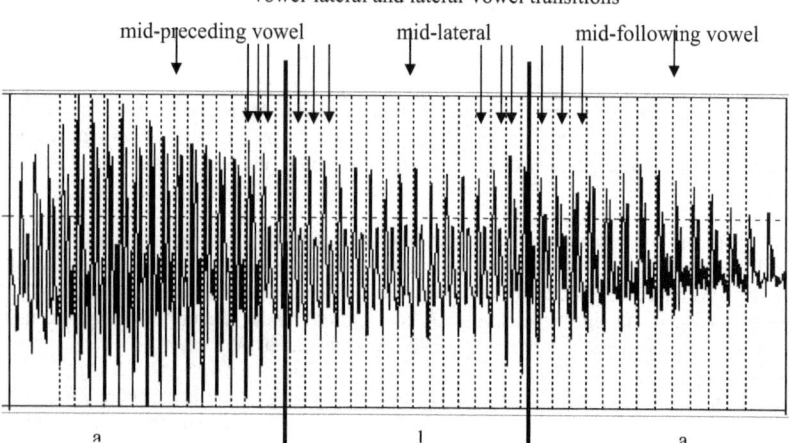

vowel-lateral and lateral-vowel transitions

mid-preceding vowel   mid-lateral   mid-following vowel

Figure 2.   Waveform with glottal closures (dashed vertical lines) for [ala], from Warlpiri *calaŋu*; spectral amplitude measurements were taken at the midpoints of the glottal pulses indicated by arrows.

Spectral amplitude was also measured at points surrounding the lateral onsets and releases as follows. After Praat calculated the edges of glottal pulses, the pulse which included portions of both the preceding vowel and the lateral was identified; the amplitude of this pulse was not measured. Instead, the amplitudes of the three pulses adjacent to this boundary pulse on either side were measured. That is, across the lateral release, amplitude measures were recorded for the final three pulses of the lateral. The next pulse, which straddles the release edge, was skipped as it represented neither lateral nor vowel uniquely; the next three pulses immediately following the beginning of the following vowel were the targets of amplitude measures as well. Six similar transition amplitude measures were recorded across the lateral onset.

Finally, the amplitude of the flanking vowels was measured. The temporal position of these measurements varied according to the length of the vowels; the target position was one which would be far enough from the lateral to be representative of the vowel rather than being part of the articulatory transition between vowel and lateral, but which would still (in a relatively long vowel) be close enough to the lateral to show some effects of coarticulation, if these are in fact present. Amplitude of a vowel preceding a lateral was thus measured at the midpoint of the glottal pulse which contained the temporal middle of the vowel, or at the midpoint of the pulse which occurred at 50 ms before the lateral – whichever pulse was closer to the beginning of the lateral. A similar measure was taken at the glottal pulse either at the middle of the following vowel or 50 ms after the end of the lateral; again, whichever pulse was closer to the end of the vowel.

In each of the locations described above, the overall spectral amplitude at that time would be measured as follows. Praat would compute a spectral slice at the desired time; this slice would then be queried for the sum of energy present at all frequencies. The output value would be given in Pascal; this was converted to decibels by the formula in (4), where *Pa* is the amplitude in Pascal and *dB* is the corresponding amplitude in decibels.

(4) $dB = 20 \times \log \left(\left(Pa \div .01 \right)^{.5} \div .00002 \right)$

These decibel values were then converted from absolute amplitudes to differences between the energy in the nearest flanking vowel and the energy at the target location; that is, the amplitude of the middle of the preceding vowel was subtracted from the each of the amplitude measures taken across the vowel-to-lateral transition, and the amplitude of the middle of the following vowel was subtracted from each of the lateral-to-vowel transition measures; the resulting normalized measurements are discussed below. Therefore, a point with a positive normalized amplitude is louder than the middle of the flanking vowel, and a negative amplitude is one which is quieter than the flanking vowel. This normalization allowed comparisons of amplitude across tokens, as it reduced token-particular variation in terms of e.g. speaker volume or recording volume. Using the nearest flanking vowel as the reference in normalization further reduced within-token amplitude variations that could result from the location of stress with respect to the lateral; this was necessary as there was not enough data available to control for stress location and so tokens varied as to whether the preceding or following vowel, or neither, was stressed.

The amplitude measure taken at the midpoint of the lateral was compared to both the preceding and following vowels, in order to calculate the similarity – and thus coarticulation – between each of these flanking vowels and the lateral itself. These two comparisons effectively served to normalize the mid-lateral amplitude measure.

3.   Results

Two major questions were investigated. First, does the preceding vowel show more coarticulation with the lateral than the following vowel, thus indicating that the lateral is are left-anchored (section 3.1)? Second, does the vocalic portion of the vowel-to-lateral transition show more acoustic characteristics of the lateral than does the vocalic portion of the lateral-to-vowel transition, further indicating that the laterals are left-anchored (section 3.2)? The basic result of these analyses is that there is no evidence of left-anchoring from the measurements taken here, in terms of either of full vowel coarticulation or of coarticulation in transitions from vowels into laterals; evidence for this conclusion follows.

3.1.   Flanking vowel coarticulation

The differences between the amplitude at the middle of the lateral and the amplitudes at the midpoints of preceding and following vowels was calculated in order to measure the extent to which laterals induce coarticulation in flanking vowels. The results are as shown in figure 3, where a short bar in the graph represents a small acoustic difference; this indicates that a large amount of coarticulation is present between vowels and laterals at the given place of articulation.

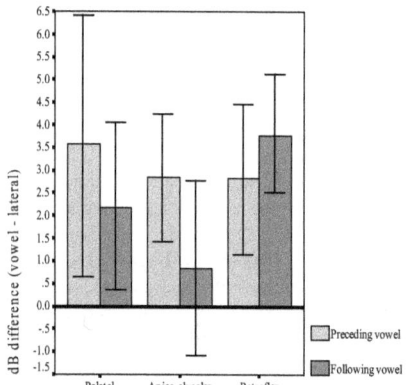

Figure 3.   Spectral amplitude differences between laterals and flanking vowels (with 95% CI), indicating flanking vowel coarticulation with laterals; low values indicate a high degree of coarticulation. Palatal data (n = 20) from Jingulu and Warlpiri; apico-alveolar (n = 32) and retroflex (n = 34) data from Jingulu, Warlpiri, and Ngandi.

The differences in vowel coarticulation in preceding versus following vowels are not significant at any place of articulation; this measure therefore fails to provide any evidence that laterals are acoustically left-anchored. We do see a pattern of coarticulation which fails to reach significance; however, this pattern also fails to suggest left anchoring, as follows.

Palatal and alveolar laterals tend to be more similar to following vowels than they are to preceding vowels, indicating that they induce more coarticulation in following vowels than in preceding vowels. Retroflex laterals are coarticulated with preceding vowels roughly to the same extent that alveolars are, though retroflexes are more different from, and thus less coarticulated with, following vowels than with preceding vowels.

The fact that retroflexes tend to induce a different pattern of coarticulation than do the other two laterals is likely related to the common claim that retroflexion is generally heard in preceding vowels but not in following vowels, and the observation that retroflexion is itself left-anchored. This suggests that this lack of coarticulation in following vowels (relative to that present in preceding vowels) reflects a general property of retroflex consonants rather than one of retroflex laterals in particular, and so this pattern is irrelevant to the question of whether laterals themselves are left-anchored.

If preceding vowels were generally more coarticulated with laterals than were following vowels, this could be evidence for left-anchoring in laterals; however, since it is in fact the following vowels which show greater coarticulation for two of the three places of articulation – and thus according to the Licensing by Cue principles themselves contain a greater degree of information than do preceding vowels about the laterals – this measure of vowel coarticulation fails to provide evidence that laterals are left-anchored.

3.2.   Lateral edge events

Left-anchoring could also be realized in the transitions between laterals and flanking vowels. If there were a significant asymmetry between the final, lateral-adjacent portion of the preceding vowel and the initial portion of the following vowel, and if the end of the preceding vowel were more lateral-like than the beginning of the following vowel, this would be strong evidence that laterals are in fact left-anchored despite the lack of left-anchoring throughout the middles of the flanking vowels. True anticipatory coarticulation which would signal left-anchoring should cause the final portion of a pre-lateral vowel to reach (or approach) an amplitude characteristic of a lateral (approximately 3 dB less than mid-vowel amplitudes, for the laterals measured here) before the beginning of the lateral. This study found no such pattern of anticipatory coarticulation.

In the amplitude graphs in figure 4, the labeled positions of the six measurements along vowel-to-lateral transitions correspond to the six labeled measurement positions along the lateral-to-vowel transitions, where the temporal order of the second series is reversed. That is, the measure taken in the preceding vowel at the third pulse before the lateral and the measure taken in the following vowel at the third pulse after the lateral are both labeled "1", as each is the outermost measurement taken along the given transition. Each subsequent pulse at which a measurement was taken across the transition is numbered sequentially, 2-6. I will use terms like "outermost" and "outside" in discussing positions in vowel-lateral-vowel sequences; these terms are generally symmetrical, referring to positions near both edges of the lateral unless otherwise specified; they take the middle of the lateral as their center and the edges of the lateral as the reference points. The "outermost" transition measurements are those furthest from the lateral within either vowel – the third pulse before the lateral onset and the third pulse after the lateral release. The "innermost" measurements are the third pulse after the lateral onset and the third before the lateral release. The heavy horizontal lines in these graphs represent the energy in the flanking vowels, to which other amplitude measures have been normalized; the dashed vertical lines represent the edges of the laterals. The dB measures displayed are differences between the amplitude at a given time and the amplitude of the nearest flanking vowel.

As figure 4 shows, transitions between laterals and flanking vowels are characterized by overall decreases in energy across the transitions at either edge of the laterals.

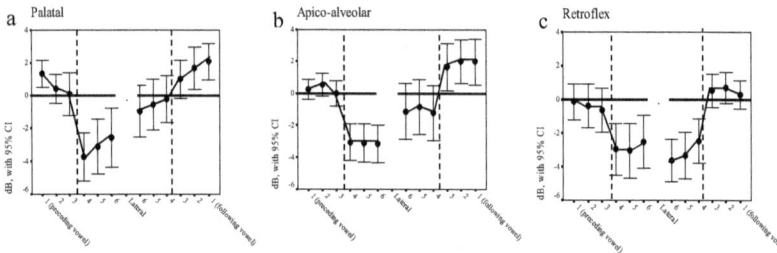

Figure 4.   Spectral amplitudes across transitions between laterals and vowels (with 95% CI), by place of articulation. Palatal data (n = 20) from Jingulu and Warlpiri; apico-alveolar (n = 32) and retroflex (n = 34) data from Jingulu, Warlpiri, and Ngandi.

These transitions are not smooth decreases from mid-vowel amplitudes (whose normalized values are represented by the heavy horizontal lines) to lateral amplitudes; rather, the amplitude immediately outside the edge of laterals (at positions 1-3) is generally greater than the mid-vowel amplitude, indicating that laterals induce amplitude increases prior to their onsets and following their releases (see section 5 for discussion of these increases). The energy peaks outside the lateral edges are asymmetrical such that the pre-lateral increase is relatively shallow compared to the increase following the end of the lateral.

Across the edges of the laterals (between positions 3 and 4, at the dashed vertical line), the energy decreases rapidly to levels below that characteristic of the vowels. The presence and magnitude of onset versus release asymmetries in energy inside the edge of the lateral varies by place of articulation. Alveolars and palatals have less energy inside the onset than inside the release; palatals also have a more abrupt energy decrease at their onset compared to at their release. Retroflexes have similar energy levels inside both edges.

In deciding whether or not these transitions are evidence for left-anchoring, only the amplitudes of the portions of vowels immediately adjacent to lateral edges (positions 1-3) are relevant. One-way ANOVA analyses which compared onset and release amplitudes across vowel-lateral transitions at each place of articulation produced no significant difference between onsets and releases for palatal and retroflex laterals (palatal: $F(1, 18) = 2.131$, $p = .162$; retroflex: $F(1, 29) < 1$, $p = .936$). There was a significant difference between onset and release amplitudes for apico-alveolar laterals ($F(1, 31) = 6.807$, $p = .014$); however, as figure 4 indicates, this change does not seem to indicate anticipatory coarticulation but rather simply a smaller energy peak preceding a lateral onset than following its release.

These data do not show a pattern which would indicate left-anchoring, as described above. Instead, the asymmetry in alveolar lateral onset and release transitions follows from the fact that, while the amplitude peak preceding the onset of alveolar laterals is relatively similar to those preceding palatal and retroflex laterals, the peak following an alveolar release is quite a bit larger than the pre-alveolar peak (and also larger than the peaks following releases of other laterals); thus there is a significant asymmetry between the transitions flanking alveolar laterals which is not present for other places of articulation.

Repeated-measure ANOVA analyses of the changes between each sequetial pair of pulses indicated that the amplitudes within the vowel (positions 1-3) and within the lateral (positions 4-6) are relatively stable and consistent – that is, there is no significant change in amplitude within these positions. The only significant amplitude change occurred between these two amplitude plateaus, where there is a significant change in amplitude across both onset and release edges of all laterals ($F(5, 11) = 39.116$, $p < .001$). These sudden transitions between otherwise stable amplitudes are another indication that laterals are not left-anchored, as there is no evidence of gradual change across the pre-lateral portion of the preceding vowel.

4.   Discussion of results

The experiment reported here has shown that laterals in Australian languages are acoustically asymmetrical in terms of their spectral amplitude; there are two primary differences in onset vs. release characteristics. First, there is a tendency for palatal and alveolar laterals to induce more coarticulation in following vowels than in preceding vowels, while retroflex laterals show the opposite pattern (figure 3). There is also a consistently greater increase in energy relative to the flanking vowel following the lateral release than there is preceding the lateral onset; this trend reaches significance in alveolar laterals (figure 4).

These asymmetries, however, do not support the hypothesis that laterals are acoustically left-anchored, and thus laterals' implicational distribution across postvocalic, word-initial, and postconsonantal environments cannot be explained based on the acoustic properties measured here. Such an account of lateral phonotactics would depend crucially on the acoustics of vowels which precede laterals showing characteristics of laterals before a lateral onset. For example, to find evidence of this asymmetry in terms of spectral amplitude, preceding vowels should be overall more coarticulated with (i.e. have amplitudes more similar to) laterals than are following vowels; for two of the three laterals analyzed here, the opposite pattern of coarticulation emerged.

Another pattern which would support a Licensing by Cue analysis would be one in which the final portion of the vowel before the lateral onset would take on acoustic characteristics of a lateral; this was also not the case. The asymmetry in transitions before and after laterals was due to a larger increase in amplitude (relative to both the lateral and also the amplitude at the midpoint of the vowel) following a lateral than preceding the lateral, rather than due to the amplitude of the preceding vowel decreasing from vowel-like values to lateral-like values before the lateral began.

As only a single aspect of lateral acoustics was measured here, it is possible that a different acoustic property might reveal just the sort of asymmetry predicted by Licensing by Cue for left-anchored segments. It is not clear at this time, though, which alternative acoustic property might reveal such an asymmetry. As discussed in section 2, formant transitions did not appear to be a locus of asymmetry, and the measure of spectral amplitude used here captures many of the spectral changes characteristic of laterals and thus seems a natural place to find any extant asymmetry. The lack of evidence for acoustic left-anchoring found here thus does not disprove the possibility of a Licensing by Cue analysis of these patterns, though it does render it increasingly unlikely. Further, this analysis did find asymmetries in spectral amplitude changes; when these asymmetries are interpreted within the Licensing by Cue framework, they make unexpected predictions regarding lateral phonotactics.

If anything, the amplitude asymmetries identified here suggest that laterals may be right-anchored according to Licensing by Cue. That is, following vowels more commonly show more coarticulation than do preceding vowels. The asymmetry in the amplitude peaks outside the lateral onset and release could also be interpreted to support this claim, as follows. Stevens and Blumstein (1978; Blumstein and Stevens 1979, 1980; also Delgutte (1997); cf. Walley and Carrell 1983) claim that the human perceptual apparatus is designed to attend to abrupt acoustic events whose frequency spectra uniquely and invariantly define phoneme identity. This claim is based on acoustic and perceptual experiments focusing on characteristics of the short-time spectra of release bursts following /b/, /d/, and /g/, which are shown by Stevens and Blumstein (1978) to differentiate these segments by place of articulation with 85% accuracy.

Following these premises, one could claim that the dramatic rises in amplitude which follow lateral releases capture the attention of the auditory system.[4] While amplitude peaks occur both before and after laterals, the latter peak should attract more auditory attention. This peak is generally larger and, more importantly, represents a much larger increase in amplitude: the amplitude rise from the lateral to the post-lateral peak is generally at least 2 dB, while the rise from the vowel to the pre-lateral peak is usually less than 1 dB. Therefore the frequency characteristics of these post-lateral peaks should play a major role in identifying laterals (and possibly differentiating among them). As this aid to lateral identification lies in the vowel following a lateral, laterals should be regarded as acoustically right-anchored, if anything.

Despite the acoustic evidence suggesting that Australian laterals are right-anchored, this acoustic property does no 'work' in the grammar of any Australian

---

[4] This claim is supported by Marlen-Wilson, Brown, and Zwitzerlood (1989).

language. That is, more information which could aid a listener in identifying a lateral lies in a post-lateral vowel than in a pre-lateral vowel. Licensing by Cue predicts that the asymmetrical presence of such information should drive a language to preferentially license e.g. laterals in contexts where this information will be available to listeners. Right-anchored segments should then be phonotactically sensitive to the segmental environment at their right edge, preferring to appear before vowels and avoiding preconsonantal positions. This, however, is never the case; laterals are never sensitive to their rightward environment, but rather are extremely restrictive in terms of which leftward contexts they will tolerate. This is evidence that these patterns should not be explained in terms of Licensing by Cue but rather using the sonority and perceptually-based analysis developed in the following section.

5.   Alternative: Formal phonotactic restrictions

The absence of acoustic evidence of left-anchoring suggests that Licensing by Cue is not responsible for the phonotactic restrictions on laterals, and that a formal account will instead appropriately describe these phenomena. Such account must describe their appearance (and lack thereof) both word-initially and also postconsonantally. This section will present first Smith's (2002) positional augmentation analysis of the word-initial restrictions, and then a syllable-contact account of the postconsonantal restrictions following Gouskova (2002, 2004). See also Flack (2004) for a more detailed discussion of these restrictions.

5.1.   Word-initial restrictions

As described above, Australian languages like Panyjima don't allow laterals to appear word-initially. This restriction is discussed by Smith (2002), who argues that word-initial onsets are preferentially of low sonority, as the acoustic properties of low-sonority segments maximize the perceptual salience of a word beginning and facilitate word recognition. She encodes this tendency in the sonority-based [*ONSET/X]/σ_1 constraint hierarchy in (5) and (6).

(5)   [*ONSET/X]/σ_1 The onset of the initial syllable in a word must have sonority less than that of X.

(6)   [*ONS/Glide]/σ_1 » [*ONS/R]/σ_1 » [*ONS/L]/σ_1 » [*ONS/N]/σ_1 » [*ONS/Stop]/σ_1

$$\uparrow$$

IDENT[F][5]

Ranking IDENT[F] below [*ONSET/Lateral]/σ_1 forces low-sonority nasals and stops to appear faithfully in word-initial position, but bans initial high-sonority glides, rhotics, and (crucially) laterals, as shown in (7).

---

[5] As this is a phonotactic effect, alternations provide no evidence about which faithfulness constraint is low-ranked and therefore fails to protect initial laterals; I use IDENT[F] here, but MAX or DEP would work as well.

(7)

/lana/	[*ONS/Lateral]/σ_1	IDENT[F]
lana	*!	
→ tana		*

Using this sonority-based hierarchy to prevent initial laterals correctly predicts that higher-sonority segments like rhotics will be banned in languages lacking initial laterals. This prediction is borne out in Australian languages; Panyjima is typical in banning its full inventory of both laterals and rhotics in initial position.[6]

(8)   <u>Panyjima</u>   (Dench 1991)   Liquids: r ɾ ̣l l ̣ʎ   Word-initial: *r *ɾ *ḷ *l *ḷ *ʎ

5.2.   Postconsonantal restrictions

Most Australian languages (e.g. Panyjima and Anindilyakwa) also prevent laterals from appearing postconsonantally. As Australian languages exhibit quite simple syllable shapes, without complex margins, this means that laterals do not appear as the second member of a medial coda-onset cluster. Evidence that this is a syllable contact effect, rather than simply a restriction on sequences of phonemes, is found in languages like Gooniyandi (McGregor 1990). Here, CL clusters are possible when they are word-initial (tautosyllabic), but not medial (heterosyllabic): *plan.pi.ra* 'on one's back', but **kap.la*. Laterals are therefore only banned as the second member of a coda-onset cluster, rather than simply in all postconsonantal positions.

In order to formalize this syllable contact restriction, we must define sonority values of laterals and other segments. Parker (2002) found that the sonority values given in (9) (the values shown are those for segments which occur in Australian languages) correlate with segments' acoustic intensity; relevant earlier work on numerical sonority values includes Clements (1990), Selkirk (1984), and Steriade (1982).

(9)

w j	11
r	10
ḷ l ḷ ʎ	9
ɾ	8
m ṇ n ṇ ɲ ŋ	6
p ṭ t ʈ c k	1

These sonority values can be used with Gouskova's (2002, 2003) *DISTANCE X constraint hierarchy. This enforces syllable contact restrictions by preferentially licensing coda-onset clusters with steeply falling sonority contours. Constraints of the form in (10)

---

[6] Factorial typology predicts that glides should occasionally fail to follow the general sonority-based phonotactic patterns that other consonants do, because glides have vocalic [V-Place] features and therefore pattern somewhat like vowels (Clements 1991). Formally, this means that the ranking IDENT[V-Place] » *sonority restrictions* » IDENT[F] is possible, and that glides may be protected in positions where other high-sonority consonants are banned. Throughout Australian languages, this is the case: glides may generally surface word-initially and postconsonantally.

are in the fixed ranking in (11), where a relatively high-ranked constraint like *DISTANCE +3 assigns violations to clusters like *m.l*, where the sonority of the onset *l* is three degrees higher than the sonority of the preceding coda *m*. Such a cluster with a steep rise in sonority is worse than a cluster like *l.m*, where the sonority of the coda is greater than that of the onset.

(10)   *DISTANCE X   The sonority difference between consonants in a heterosyllabic cluster may not be X.[7]

(11)   ← steep rise ————————————— flat ——————————— steep fall →
         ...*DIST +3 » *DIST +2 » *DIST +1 » *DIST 0 » *DIST -1 » *DIST -2 » *DIST -3 ...
         e.g. *m.l   *ʈ.c   *l.m

The *DISTANCE constraints relevant to possible clusters in Australian languages are given in (12). Ranking *DISTANCE -2 » IDENT[F] » *DISTANCE -5 correctly predicts a lack of lateral-final clusters; this ranking can be determined based on the inventory of clusters in Australian languages, as will be described below.

(12)   *DIST +8 » *DIST +3 » *DIST +1 » *DIST 0 » *DIST -1 » *DIST -2 » IDENT[F] » *DIST -5
        e.g. *p.l   *m.l   *r.l   *ʎ.l   *r.l   *w.l   ✓n.t

IDENT[F] must outrank *DIST -5, because of the possible clusters in Panyjima (whose cluster inventory is typical of Australian languages), the worst cluster – i.e. the cluster with the smallest sonority drop – is a nasal followed by a stop, as in *kan.ta* 'leave it!' The nasal has a sonority value of 6, and the stop has a sonority value of 1; the slope is therefore -5, so IDENT[F] » *DISTANCE -5 will allow this cluster to surface faithfully.

(13)

/kan.ta/	IDENT[F]	*DIST -5
→ kan.ta		*
kal.ta	*!	

*DIST -4 must outrank IDENT[F], because clusters with a sonority drop of -4 (e.g. *r.ŋ*) never surface faithfully.

(14)

/tarŋa/	*DIST -4	IDENT[F]	*DIST -5
tar.ŋa	*!		
→ tar.ka		*	

Given this ranking *DIST -4 » IDENT[F] » *DIST -5, and the rest of the *DISTANCE hierarchy, the absence of lateral-final clusters is predicted as shown in (12). The best possible lateral-final cluster – i.e. the one with the maximal sonority drop – would be composed of a glide followed by a lateral. This would violate *DISTANCE -2, which

---

[7] The *DISTANCE hierarchy is formally developed through relational alignment of the *ONSET/X and *μ/X hierarchies, which prefer low-sonority onsets and high-sonority codas, respectively; this produces a syllable-contact hierarchy which may also be characterized in terms of numeric sonority contours.

outranks *DISTANCE -4 and therefore outranks IDENT[F]. This and all other lateral-final clusters (which would have even less optimal sonority contours) are therefore banned.

(15)

/tawla/	*DIST -2	*DIST -4	IDENT[F]	*DIST -5
taw.la	*!			
➜ taw.ta			*	

5.3.   The implicational nature of lateral phonotactics

I have shown that the common word-initial and postconsonantal bans on laterals in Australian languages can be explained in terms of laterals' relatively high sonority, as these phonotactic positions prefer to host low-sonority segments. The restrictions are formally encoded in sonority-based constraint hierarchies targeting initial and postconsonantal onsets; the rankings of these constraints necessary to predict the restrictions on laterals also correctly predict other phonotactic patterns found in these languages.

As noted earlier, lateral phonotactics are implicational: if laterals are licensed word-initially in a language, they also appear postvocalically; if they are licensed postconsonantally, they will be licensed word-initially as well. An analysis of lateral phonotactics based on their acoustic properties, and formalized using Licensing by Cue, would inherently capture the implicational nature of the phonotactics. As shown above, however, lateral acoustics do not support such an acoustically-based analysis. Instead, as shown in this section, the lateral restrictions can be explained using the *DISTANCE and [*ONSET/X]/σ_1 constraint hierarchies.

The use of independent constraints does not inherently capture the implicational nature of the restrictions, though; it should be possible for a language to ban laterals word-initially but allow them to surface postconsonantally. Given the crosslinguistic tendencies to impose sonority-based restrictions word-initially and postconsonantally, however, as well as the tendency towards phonological consistency within Australian languages, this accidental implicational relationship is nevertheless expected, as follows.

A small set of languages ban initial high-sonority segments: the Iglesias dialect of Campidanian Sardinian (Bolognesi 1998) and Mbabaram (Dixon 1991) ban initial rhotics, and Mongolian (Poppe 1970), Kuman (Lynch 1983), and most Australian languages (Hamilton 1996) ban all initial liquids. While such high-sonority segments are nonoptimal from a perceptual perspective in word-initial position, most of the world's languages allow them. This means that there is a strong tendency for [*ONSET/Lateral]/σ_1 to be relatively low-ranked. In contrast, syllable-contact restrictions against coda-onset clusters without sharp sonority drops are prevalent (see Gouskova (2003) and references cited therein); this indicates that *DISTANCE constraints (which tend to prevent lateral-final clusters) tend to be relatively high-ranked cross-linguistically.

A likely functional motivation for this bias arises in the competing pressures to which word-initial position is subject. Initial segments are preferentially of low sonority in order to facilitate identification of word boundaries. However, in order to facilitate rapid word recognition, it is also important to maximize the number of segmental contrasts realized in initial position (this tendency is captured in positional faithfulness constraints; see e.g. Beckman 1998). The cross-linguistic desirability of low-sonority initial segments is often sacrificed in order to maximize initial contrasts. No similar competing pressures are relevant in medial clusters, and thus syllable-contact restrictions may apply more freely across languages.

Given these tendencies, it is not surprising to find that *DISTANCE constraints apply more frequently across Australian languages than do [*ONSET/X]/σ_1 constraints. The fact that these languages go beyond a mere tendency and are deeply consistent in allowing [*ONSET/X]/σ_1 constraints to be active (i.e. high-ranked) only if those *DISTANCE constraints which ban postconsonantaly laterals are highly-ranked as well is also unsurprising, given that the phonological systems of Australian languages tend to be quite consistent (Hamilton 1996, Dixon 2002). Lateral restrictions are only one of many phonological patterns which are common across Australian languages. While individual languages vary in where laterals may appear, they adhere to a basic pattern where laterals are dispreferred word-initially, and even more strongly dispreferred postconsonantally.

6.   Conclusion

The specific goal of this paper was to evaluate two possible frameworks which might provide accounts of lateral phonotactics in Australian languages; a broader goal was to examine the roles of acoustic and formal explanations in phonology. Towards the first goal, I have demonstrated that the acoustics of laterals do not explain their phonotactic distribution; that is, the Licensing by Cue framework cannot explain the contextual restrictions on laterals. I have further shown that it is possible to explain these restrictions in terms of formal constraints. As there is no similarly possible acoustic analysis, this formal analysis appears to be correct.

The study of lateral acoustics has also contributed to an understanding of the relative explanatory power of formal and acoustic frameworks. Previous work on Licensing by Cue has shown that this framework can explain phenomena like retroflex licensing, where segmental acoustics and phonotactics (as well as patterns of assimilation in clusters) conspire to indicate that segments are e.g. left-anchored. In such phenomena, phonotactic patterns seem to be the natural result of segmental acoustics, and acoustics are the basis of the constraints that govern phonotactics in this situation.

A strong version of Licensing by Cue could claim that all such patterns of edge-sensitive phonotactics are the result of acoustic asymmetries, and that any acoustically-asymmetric segments would show corresponding phonotactic asymmetries. The acoustic properties of laterals, however, demonstrate that this strong claim cannot be true. Lateral phonotactics are describable in terms of their left-edge segmental context; lateral acoustics, however, are if anything right-anchored. This mismatch, where lateral phonotactics do not follow from their acoustics and instead must be the result of formal, sonority-based principles, demonstrates that Licensing by Cue cannot be strictly deterministic but must instead be one of many mechanisms operating in a phonological grammar to determine phonotactic patterns. While grammars may allow segmental acoustics to determine phonotactics, they may also allow other (formal) properties of segments, like sonority, to determine phonotactics as well.

Kathryn Flack

Appendix.   Tokens

Tokens used in the acoustic experiment. Underlined a<u>L</u>a sequences are those which were measured; a number following a token indicates multiple instances of that word.

Jingulu	*Warlpiri*	*Ngandi*

<u>Palatal</u> (5)

caʎanana

caʎaŋkunu

kaʎaɽa-aʈi

kaʎaɽu-wuɽu

naʎaɾiɲincu

<u>Palatal</u> (15)

cama<u>ʎa</u>ḻa

cama<u>ʎa</u> (2)

wa<u>ʎa</u>ŋka (10)

wa<u>ʎa</u> (2)

<u>Alveolar</u> (9)

taʈuw<u>ala</u>

ka<u>la</u>ɽa

w<u>ala</u>nca (3)

want<u>ala</u>-alu

wup<u>ala</u>

jaŋp<u>ala</u>-nu

jaɾintu-w<u>ala</u>

<u>Alveolar</u> (13)

ca<u>la</u>ɲu (6)

caɲ<u>ala</u> (4)

naɲ<u>ala</u> (3)

<u>Alveolar</u> (10)

abunk<u>ala</u>lakgalala

abunkal<u>ala</u>kgalala

abunkalalakg<u>ala</u>la

aḍapb<u>ala</u>ɲ

am<u>ala</u>pinybiɲ

kug<u>ala</u>ɾ

mab<u>ala</u>ɾa?

maḍ<u>ala</u>wuʈbuʈ

magal<u>ala</u>n? (2)

<u>Retroflex</u> (8)

ma<u>lạ</u>ḷuka

wa<u>lạ</u>ku-ɳi (4)

wa<u>lạ</u>ku (3)

<u>Retroflex</u> (12)

ɳa<u>lạ</u>ɾimi (10)

ɲaɾpaɻa<u>lạ</u> (2)

<u>Retroflex</u> (14)

aga<u>lạ</u>ḷga<u>lạ</u>ḷ (3)

agal<u>ạ</u>ḷgal<u>ạ</u>ḷ

am<u>ạ</u>ḷa

a<u>lạ</u>puɳiɲ (2)

kuc<u>ạ</u>ḷa

kuwa<u>lạ</u>n?

magar<u>ạ</u>ḷa<u>lạ</u>an

magaral<u>ạ</u>ḷ<u>ạ</u>ḷan

naɲa<u>lạ</u>ɳci (3)

References

Beckman, Jill (1998) Positional Faithfulness. Doctoral dissertation, University of Massachusetts, Amherst.

Blumstein, Sheila E., and Kenneth N. Stevens (1979) Acoustic invariance in speech production: Evidence from measurements of the spectral characteristics of stop consonants. *Journal of the Acoustic Society of America* 66: 1001-1017.

Blumstein, Sheila E., and Kenneth N. Stevens (1980) Perceptual invariance and onset spectra for stop consonants in different vowel environments. *Journal of the Acoustical Society of America* 67: 648-662.

Boersma, Paul, and David Weenink (1992) *Praat: Doing Phonetics by Computer.* [www.praat.org]

Bolognesi, Roberto (1998) *The Phonology of Campidanian Sardinian: A Unitary Account of a Self-Organizing Structure.* Amsterdam: Holland Institute of Generative Linguistics.

Clements, G. N. (1990) The role of the sonority cycle in core syllabification. In J. Kingston and M. Beckman (eds.) *Papers in Laboratory Phonology I: Between the Grammar and the Physics of Speech.* Cambridge: Cambridge University Press.

Clements, G. N. (1991) Place of articulation in consonants and vowels: A unified theory. In *Working Papers of the Cornell Phonetics Laboratory*, vol. 5, pp. 77-123. Ithaca: Cornell University.

Delgutte, Bertrand (1997) Auditory neural processing of speech. In W.J. Hardcastle and J. Laver (eds.) *The Handbook of Phonetic Sciences*, pp. 507-538. Oxford: Blackwell.

Dench, Alan (1991) Panyjima. In R.M.W. Dixon and B.J. Blake (eds.) *Handbook of Australian Languages*, vol. 4, pp. 125-244. Oxford: Oxford University Press.

Dixon, R. M. W. (1991) Mbabaram. In R.M.W. Dixon and B.J. Blake (eds.) *Handbook of Australian Languages,* vol. 4, pp. 348-402. Oxford: Oxford University Press.

Dixon, R. M. W. (2002) *Australian Languages.* Cambridge: Cambridge University Press.

Flack, Kathryn (2004) Lateral phonotactics in Australian languages. Ms. University of Massachusetts, Amherst.

Gerfen, Chip (2001) A critical view of licensing by cue: Codas and obstruents in Eastern Andalusian Spanish. In L. Lombardi (ed.) *Segmental Phonology in Optimality Theory*, pp. 183-205. Cambridge: Cambridge University Press.

Gouskova, Maria (2002) Exceptions to sonority distance generalizations. *CLS 38: Main Session.*

Gouskova, Maria (2003) Relational hierarchies in OT. Ms. Rutgers University.

Hamilton, Philip J. (1996) *Phonetic Constraints and Markedness in the Phonotactics of Australian Aboriginal Languages.* Doctoral dissertation, University of Toronto.

Heath, Jeffrey (1978) *Ngandi Grammar, Texts, and Dictionary.* AIAS: Canberra.

Howe, Darin, and Douglas Pulleyblank (2001) Patterns and timing of glottalization. *Phonology* 18: 45-80.

Kingston, John (2002) Keeping and losing contrasts. *BLS 28.*

Ladefoged, Peter, and Ian Maddieson (1996) *The Sounds of the World's Languages.* Oxford: Blackwell.

Laughren, Mary, Robert Hoogenraad, Kenneth Hale, and Robin Japanangka Granites (1996) *A Learner's Guide to Warlpiri: Tape Course for Warlpiri (Wangkamirlipa Warlpirilki).* Alice Springs: IAD Press.

Leeding, Velma Joan (1989) *Anindilyakwa Phonology and Morphology.* Doctoral dissertation, Sydney University.

Lynch, J. (1983) On the Kuman "liquids". *Languages and Linguistics in Melanesia* 14: 98–112.

Marlen-Wilson, William D., Colin M. Brown, and Pienie Zwitserlood (1989) Spoken word-recognition: Early activation of multiple semantic codes. Ms., Max Planck Institut.

McGregor, William (1990) *A Functional Grammar of Gooniyandi.* Amsterdam: John Benjamins.

Nash, David (1986) *Topics in Warlpiri Grammar.* Garland Publishing Inc.: New York.

Parker, Stephen G. (2002) *Quantifying the Sonority Hierarchy.* Doctoral dissertation, University of Massachusetts, Amherst.

Pensalfini, Robert J. (1997) *Jingulu Grammar, Dictionary, and Texts.* Doctoral dissertation, MIT.

Poppe, Nikolaus (1970) *Mongolian Language Handbook.* Washington: Center for Applied Linguistics.

Prince, Alan, and Paul Smolensky (1993) *Optimality Theory: Constraint Interaction in Generative Grammar.* RUCCS Technical Report no. 2.

Selkirk, Elisabeth O. (1984) On the major class features and syllable theory. In M. Aronoff and R.T. Oehrle (eds.) *Language Sound Structure: Studies in Phonology Presented to Morris Halle by his Teacher and Students*, pp. 107-113. Cambridge, MA: MIT Press.

Smith, Jennifer L. (2002) *Phonological Augmentation in Prominent Positions.* Doctoral dissertation, University of Massachusetts, Amherst.

Steriade, Donca (1982) *Greek Prosodies and the Nature of Syllabification.* Doctoral dissertation, MIT.

Steriade, Donca (1997) Phonetics in phonology: The case of laryngeal neutralization. Ms., MIT.

Steriade, Donca (1999) Alternatives to the syllabic interpretation of consonantal phonotactics. In O. Fujimura, B. Joseph, and B. Palek (eds.) *Proceedings of the 1998 Linguistics and Phonetics Conference*, pp. 205-242. Prague: The Karolinum Press.

Steriade, Donca (2001) The phonology of perceptibility effects: The P-map and its consequences for constraint organization. Ms., UCLA.

Stevens, Kenneth N. (1998) *Acoustic Phonetics.* Cambridge, MA: MIT Press.

Stevens, Kenneth N., and Sheila E. Blumstein (1978) Invariant cues for place of articulation in stop consonants. *Journal of the Acoustical Society of America* 64: 1358-1368.

Walley, Amanda C., and Thomas D. Carrell (1983) Onset spectra and formant transitions in the adult's and child's perception of place of articulation in stop consonants. *Journal of the Acoustical Society of America* 73: 1011-1022.

Department of Linguistics
South College
University of Massachusetts
Amherst, MA 01003

flack@linguist.umass.edu

The French stratum of English

Ben Gelbart

University of Massachusetts, Amherst

1.   Introduction

Every language has loan words. Quite often, one particular language may serve as the source language for a significant number of loan words in a given target language. It is not unusual that this source language has a different phonology and morphology than the target language. Thus, it is not uncommon that a subset of words in a language that are loan words share a single source language also share phonological characteristics and phonological and morphological processes to the exclusion of the rest of the lexicon.

In a number of such cases, a grammatical distinction has been claimed between native and foreign phonologies within a single language. These include Japanese (Ito & Mester 1998), Russian (Saciuk 1969, Holden 1976), and Modern Hebrew (Schwarzwald 1998). In such cases, the stratified nature of the phonological lexicon is systematic and extensive. Each stratum has certain phonological characteristics associated with it. These may include unique segments that appear only in a certain stratum, phonotactic restrictions, even unique phonological and morphological processes. Each stratum functions as a "co-grammar" while also sharing general properties with the other strata. For such languages, the stratification of the lexicon and manipulation of co-grammars is not optional behavior chosen by only some speakers. To know the language means knowing the co-grammars and co-lexicons of all the strata.

No claim is being made that by exhibiting the ability to manipulate co-grammars, speakers are displaying knowledge of the history of their language. Yet the term "foreign" is still appropriate for a number of reasons. All such cases of robust manipulation of co-grammars involve a lexicon that is stratified along lines of historical origin. While it is not necessary for speakers to know this fact in order to correctly manipulate the co-grammars, they are usually aware of it as well.

Kathryn Flack and Shigeto Kawahara (eds.), UMOP 31, 59-85.

For example, a speaker of Modern Hebrew doesn't need to know the actual historical origin of a word like *mikroskóp* to know not to shift its stress to the suffix *im* when deriving the plural *mikroskópim* (the stress of nearly all native Hebrew words is final). The tell-tale signs are that *mikroskóp* contains more than three root consonants and ends in the plosive *p*, neither of which can happen in a native word. Words that are less obviously of foreign origin like *salát*, may become indistinguishable from words of the native stratum and be pluralized as if they were native, e.g. *salatím* (Becker 2003). However, Hebrew speakers are probably in any case aware that *mikroskóp* and *salát* are not native words regardless of phonotactics and most speakers usually say *salátim* despite the lack of phonotactic cues to foreign stratal affiliation. By comparison, a word speakers know is native such as *šarát* "janitor," is always pluralized with a shift of stress: *šaratím* and never **šarátim*.

To argue that data of these kinds are evidence that foreignness plays a role in phonology is to assume that identical speech sounds are being treated differently by phonological processes or constraints because speakers know that words belong to different word classes. Such an assumption suggests a testable prediction, namely that physically identical speech sounds might be perceived differently depending on the foreignness of their context. Any evidence of such a perceptual bias would confirm that foreignness plays a role in phonology and would make alternative explanations for the attested production patterns superfluous.

The claim that foreignness plays a role in phonology has been questioned, for example, by Inkelas, Orgun and Zoll (1997) who account for the same sorts of phenomena by enriching the representation that underlying forms can contain. Inkelas et al. reduce the exceptional properties and behaviors of the foreign stratum to differences in underlying forms, obviating the need for separate strata.

The stratified lexicon model does not necessarily refute the Inkelas et al. approach of enriching underlying representation to account for differing behaviors. In the case discussed by Inkelas et al. voiced obstruents are devoiced in certain environments in native Turkish roots but not in loaned roots. Inkelas et al. propose that there is simply a three-way contrast in Turkish obstruents between those pre-specified for voicing (either voiced or voiceless) that do not alternate, and a third, underspecified category whose voicing is determined by its environment. It is possible that Turkish really does represent the segment *d* two different ways underlyingly: one underspecified *d* that devoices in certain environments and one fully specified *d* that never devoices. The fact that the fully specified *d* only occurs in words of foreign origin may not necessarily be of interest to speakers. As Inkelas et al. point out, lexical subclasses can be arbitrarily carved out of any lexicon by grouping together all the words that contain or don't contain some feature X. It is unlikely that speakers compartmentalize their lexicon to such an extent, and in any case doing so would not make a difference in their linguistic behavior and so it is difficult to determine whether they do or not. The only cases in which we can test speakers' awareness to such lexical subclasses are when they involve correlations of two or more features, say, if words that contain feature X are thought more likely to contain feature Y and less likely to contain feature Z. It is by testing awareness of the correlations of marked phenomena over the lexicon that we can argue for the existence of lexical strata.

There are a number of arguments for favoring approaches like the stratified lexicon that appeal to foreignness as a primitive of phonological theory. In these theories, there is something special about foreign features: they are the result of diacritic markings on phonological rules, or highly ranked duplicate versions of faithfulness constraints. By comparison, the strata-neutral approach loses the distinction between regular and exceptional forms, leaving it to coincidence that one type is more prevalent and less restricted than the other. There is also the historical process of regularization that assimilates words from the exceptional class to the non-exceptional and not the other way around. However, one of the most forceful arguments for the role of foreignness in phonology is evidence that speakers can appeal to it directly when making linguistic decisions.

In speech perception tests, listeners are asked to judge a speech sound in the context of either foreign or native material. The null hypothesis is that the context (which is held constant for other factors) will not influence the judgment. The alternative hypothesis is that the perception of speech sounds is in fact influenced by whether listeners perceive a foreign or native context. If speakers tend to choose foreign alternatives in foreign contexts more than in neutral contexts, then they are treating foreignness as a category, or least a recognizable correlation of features. Such evidence is not supported by strata-neutral theories like that of Inkelas et al. for whom "foreign" properties and behaviors are unrelated properties of possible underlying forms.

By comparison, mere generalizations in production data are subject to the suspicion that they play no role in linguistic knowledge. For instance, before the borrowing of loanwords from Norse, no words in the English language began with *sk-*. Later borrowings from Latin added some non-Norse *sk-* initial words, but it is probably still true that most *sk-* initial words in English are of Norse origin. A question that often arises in the discussion of such phenomena is "Who cares?" Such historical curiosities are of interest to generative linguists only when there is reason to believe that speakers are incorporating such knowledge into their grammar, or at least are aware of such knowledge to the extent that they use it to make linguistics decisions. The above-mentioned case of *sk*-initial words in English almost certainly plays no such role. The property of beginning with *sk-* is not correlated with any other phonotactic or morphological property in English. There is no reason to assume that speakers of English treat *sk-* initial words any differently than, for example, *ta-* initial words.

The test cases are those where two or more phonotactic cues are thought to be associated with a single lexical stratum. Japanese, with its well documented stratified phonological lexicon, provides an obvious test case. An experiment by Moreton and Amano (1999)[1] showed that Japanese listeners significantly shifted the perceptual boundary between word-final long *a:* and short *a* in nonce words depending on how phonologically foreign the rest of the word is. Moreton and Amano used nonsense words that contained phonological cues to particular stratal affiliations. In Japanese, only the Foreign stratum can contain both long and short *a* word-finally. Other strata can contain short *a*, but not long *a:* in word-final position. Some of the carrier nonsense words

---

[1] The same experiment is also reported as Experiment 6 in Chapter 4 of Moreton (2000).

contained the consonants *p* and *f*, which are only found in the Foreign stratum of Japanese. Other nonwords contained only consonants found in all strata. Still other nonwords contained r^y and h^y, which are only found in the Sino-Japanese stratum. The perceptual boundary between short and long *a* was shown to be shortest when the carrier nonword contained Foreign cues, longer when it contained neutral cues, and longest when the word contained Sino-Japanese cues. This suggests that listeners are more likely to identify word-final *a* as long in those strata where it is phonotactically legal and are less likely to do so in those strata where it is phonotactically illegal.

The contention of this paper is that the stratified lexicon is part of linguistic behavior and therefore should be in evidence in other languages as well. English has a very large number of loanwords from a diverse range of sources including Latin, Old Norse, French, and Spanish, among the more common sources. A rather complex stratified lexicon could be constructed to include all or some of these different sources, each with its own unique phonotactic and morphological attributes. However, for the purpose of testing a perceptual bias, we will focus here on the most recent and least assimilated stratum of loans into English. Many of the lexical items in this stratum are French, and even if some of them are not, most of what can be considered the unique attributes of this stratum are French in origin.

Data from Janda, Joseph and Jacobs (1994) suggest a number of attributes that speakers of English consider to be noticeably foreign. These are summarized in Table I.

Attribute	Examples
ž	beige, mirage
Final stress in nouns	police, canoe
The suffixes *–ette, -elle, -age, -ie(r)*	cigarette, croupier, garage
No final codas (except in the suffixes *–ette*, etc.):	coup de gra(s), fleur de li(s)
Preference of *a* over other vowels, especially *æ*:	pasta, chianti

Table I. Phonological attributes of the French stratum of English.

This stratum is termed "French" here because its attributes are all present in the French language and more crucially, are familiar to English speakers by means of contact with French loanwords. However, here the French connection ends. Once borrowed, French loanwords into English do not cause English speakers to be aware of French phonology as a whole, but only of those phonological attributes which make these loanwords stand out among native English words. Even in these loanwords, many attributes of the actual French pronunciations have already been nativized, such as front rounded vowels. After they have been borrowed, loanwords may then undergo further nativization, causing them to lose any attributes that would have made them recognizable as loanwords. Furthermore, the complete suite of French attributes may be extended to words that in their original borrowed form contain only some of these attributes. Finally, such extension of attributes may occur in loanwords whose origin is not even French. However impurely French this stratum may be, these inconsistencies are most probably typical of foreign strata in many languages where one particular language is the source of a large number of easy to recognize loans. The Sino-Japanese stratum of Japanese does not accurately represent the phonology of Chinese either.

The process of "hyperforeignization" documented by Janda et al. (1994) suggests that these French attributes are associated with productive rules in English that are applied to novel material. The examples they cite include: "the Republic of Belaru(s)," "[sometime Israeli cabinet minister] Shimon Perés," and the pronunciation of "lingerie" as *lànžeréi*. What makes processes like these different from native processes is their scope of application. They apply exclusively to material that is perceived as foreign. A word might be perceived as foreign for one of two reasons. It may already contain at least one recognizably foreign attribute, or it may have been presented to the listener in some kind of foreign context.

In cases like *Belaru* and *Perés*, the decision to apply foreignizing processes is made on a lexical basis. These words are known to be foreign names, therefore they must conform to "foreign" phonology. It is even possible that the foreignizing occurred first at the perceptual level and that the pronunciations *Belaru* and *Perés* are faithful to the way the speaker first perceived them. In any case, there is nothing about the words that would make the pronunciations *Belarus* and *Péres* recognizably foreign on phonotactic grounds. If they had been presented as Bellaroose and Perris in a non-foreign context, there is no reason English speakers couldn't have pronounced them as *Belarus* and *Péres*.

In cases like *lingerie*, as pronounced *lànžeréi*, the actual French pronunciation *linžerí* does contain recognizably foreign attributes: the phoneme *ž* and final stress. These already existing phonotactic attributes are most probably the trigger for the change in vowel quality. There is no obvious reason why a word that means what lingerie means can't be a native word.

While the foreign context cue (especially in foreign names) is probably responsible for more cases of hyperforeignism in speech production, the phonotactic cue is the one that can be more precisely controlled in laboratory conditions.

2.   Experiment

The experiment described here is an attempt to use English to replicate the results of Moreton and Amano (1999) by relying on the phonotactic cues of the proposed French stratum of English.

Of the five attributes of the French stratum of English given in Table I, not all can serve equally well as triggers and measures. One necessary property of a measure is that it can be presented in gradient steps, so that a perceptual boundary can be measured. It would be nearly impossible to present stress gradiently, therefore final stress can not be a measure. Presence or lack of final coda, on the other hand, would probably make a good measure, but a bad trigger. Presence of a final coda could be presented gradiently: from a final vowel that fades gracefully to a clearly released burst. A speaker of English might be less likely to hear the final stop if the word contains other foreign attributes. However, definite presence or absence of a final stop might not trigger the listener's hearing other foreign attributes. This is because lack of final coda is an attribute of the French stratum, while presence of final coda is not an attribute of the English stratum. Additionally, not all triggers can be used to trigger all measures. The suffixes *-ette, -elle, -age* are exceptions to the no final coda attribute. The result is that final coda can not be used to measure the effect of the suffixes.

Moreton and Amano's (1999) design was elegant in that the triggers were used additively: there was a three-way contrast between no cues, one cue, and two cues. The result was that the measure was affected in an additive fashion. This can be done with the French stratum of English as well. For instance, the combination of final stress and *ž* might be a more powerful trigger than either one by itself. Such a two-factor contrast can be presented by using two pairs that use the same base nonsense word. One pair always has initial stress and the other always has final stress and both pairs have variants with and without *ž*.

It is also possible to combine final stress with the suffixes *–ette, -elle, -age*. Instead of simply replacing the suffix (say, replacing *–ette* with *–ek*, one could also retract the stress to an early syllable. This may have an additive effect since presumably the suffixes will not be recognized as robustly if they are not stressed. Not all cues can be combined, however. The suffixes can not be combined with absence of final codas, since they all have final codas.

Final stress is problematic as a universal cue to the French stratum because it only applies to nouns, as effectively demonstrated in a series of Experiments by Kelly (1988b). This pattern can easily be seen in the case of noun/verb near homophone pairs that differ only in stress such as *permit, extract, recall*. There are true homophones like *merit* and *debate*, but there are no near homophones where the noun is iambic and the verb trochaic. Kelly (1988b) has shown that disyllabic nonce words spoken in isolation are assumed to be verbs more often if they are iambic than trochaic. Kelly (1988a) demonstrated that written disyllabic English nonce words presented in a verb context are assigned Iambic stress more often than Trochaic stress. Thus, by using stress as a stratal cue, we run the risk of biasing the listener towards nativeness in disyllabic words no matter how the stress is placed. Initial stress might suggest a native noun, but final stress might suggest a native verb.

How can we bias listeners into assuming that nonsense words presented in isolation are nouns? One way is to always use stress in conjunction with recognizable stress-bearing suffixes like *–ette, -elle, -age* that are highly suggestive of nouns. In these cases, the suffix is not being used as a trigger, since it is present in both end points. The suffix is being used as a bolster and a block to the effect of stress placement. It is a bolster to encourage the listener to interpret the final stress variant as foreign. It is a block to keep the final stress variant from being interpreted as a verb. If we pursue this strategy, we can not test the additive effect of stress and French suffixes.

Another approach is to use trisyllabic words. As Kelly (1988a) points out, trisyllabic stress patterns can also be associated with either nouns or verbs. In particular, trisyllabic words with initial stress will tend to have secondary stress on the third syllable if they are verbs. This pattern can be seen in the noun and verb variants of near homophones like *delegate* and *alternate*. Also, trisyllabic words with penultimate stress are more likely to be nouns. Presumably, the noun/verb bias among trisyllabic words is in any case weaker than the bias among disyllabic words. Also, there are few trisyllabic nouns or verbs that have ultimate stress so this would probably make a good foreignness marker regardless of whether the listener interprets the word as a noun or a verb.

2.1. Design

The many asymmetries in the attributes of the French stratum suggest that it is better not to attempt a design as tightly controlled as Moreton and Amano (1999). It is perhaps better to use a dozen or so word pairs that are varied in their triggers and other attributes, but not necessarily in a symmetrical way. In each pair, however, the trigger (or combination of triggers) is predicted to shift the boundary of the measure in a particular direction. We can thus run planned matched-pair one-tailed t-tests on the results of each pair separately. In the case of the paired pairs, where two triggers are being tested for their additive effect, a two-way ANOVA will be used instead of a t-test. There is no a priori way to test the differences between the various pairs such as for instance, which triggers or measures lead to more robust results. We can attempt some kind of post hoc test to test any patterns that seem to emerge between pairs.

Measure Trigger	*ž*	final coda	*a*
ž only	-	iii	v
suffix only	ii	-	
stress only	i		
stress w/suffix	iv	-	
spelling only		xi	x
suffix and spelling		-	viii-ix
a & *ž*	-	xii-xiii	-
a & stress w/suffix	vi-vii	-	-

Table II. Triggers and measures. The Roman numeral refers to the nonce word pair in Table IV that represents the trigger-measure combination in question. A dash represents a trigger-measure combination that can not be tested. Blank cells are combinations that were not used due to time considerations.

In designing labels for listeners to use in registering their decisions regarding the measures, a question arises concerning orthography. English spelling is notoriously unreliable. First, it is probably a good idea to embolden the letters that spell the measure in order to focus the listeners' attention on the choice to be made and so facilitate a speedy response. The greater problem is how to spell the labels of the measure decisively so that it is clear to the listener what the choice is and it is clear to the experimenter what choice has been made. As Venezky (1970), from whom are gleaned most of the spelling generalizations used in the construction of labels in this experiment, has pointed out (p.101): "[English] *o* corresponds to seventeen different sounds, *a* to ten, *e* to nine, and the combined group [of vowel spellings] to forty-eight."

Final coda is relatively easy to spell: either the final coda is present or it isn't. It is harder to find a way to spell the difference between back *a* and front *æ*. One strategy is to double the following consonant to suggest *æ*, although word-finally it may be simpler to leave a single consonant. Use of "ah" may be a good way to suggest *a* as in "a-hah." It is also less than straightforward to consistently represent a non-native phone like *ž* for which no English letter or sequence of letters exists. While *ž* can be spelled a great many ways, as can be observed in "regime, equation, luxury, bijou, Asia," it is probably most consistently spelled as "ge." The affricate *ĵ* was chosen as the opposite endpoint to *ž*, since that is how *ž* is most often nativized in English. *ĵ* can be reliably spelled "j", except word-finally where "j" is never found in English orthography (Venezky 1970, p. 75). "j" is not found in clusters either, and so can not be duplicated to indicate that the preceding "a" is pronounced *a* and not *æ*. In these two cases, *j* must be spelled "dge." Even so, some of these spellings, when seen for the first time, may not immediately suggest to the listener the pronunciations we intend.

It is clear that some kind of training pre-test is necessary to make sure that listeners understand what the contrasts are that they are being ask to differentiate. One possible solution is to train listeners on pairs of geometrical shapes instead of graphemes. However, word pairs may differ in length and prosodic shape and where they contain the measure within the word. Listeners are also being asked to judge three separate contrasts. This may make the task more difficult than necessary. Using some kind of alphabetic representation of the entire word is thus necessary to cue listeners to both the position and the contrast that they are being asked to judge. This allows the listener to focus as much attention as possible on the task of judging the contrast.

For the measures, the labels need to be as unambiguous as possible, so we can be sure we know what decision listeners mean to convey by choosing one label over another. In spelling the rest of the word, we can allow a little more flexibility. The trigger, after all, will be unambiguously present in the sound stimulus. The only thing we need to worry about is that nothing in the spelling in the rest of the word appears to be of a different stratum than the trigger. Sometimes this will mean changing the spelling between the two trigger conditions. For instance, when the trigger is *ž*, the spelling between the two endpoints will in any case have to be different. Thus, for at least some of the word pairs, there is no other option than to use four unique spellings: one per each measure endpoint per each trigger condition.

In the word pairs where the trigger is stress placement bolstered by a French suffix, it might be necessary to use four unique spellings as well. Consider the word pair $Z_i k_i l \sim Z_i k \varepsilon l$, where the trigger is the stress and the measure is Z, a voiced obstruent ranging between *ĵ* and *ž*. When the stress is final, the spellings will have to be something like *juquelle ~ geuquelle*. However, when the stress is initial, it might be best not to spell out the French suffix *-elle*, since this suffix is always stressed and its spelling may conflict with what should be unambiguous initial stress. It might be better to use nativized spellings like *jukle ~ geukle* or even *joockle ~ geoockle*.

There are four unique endpoints in each word pair and using different spellings for each could facilitate ease of response. Also, using four spellings for all word pairs, whether necessary or not, makes the experimental design the same in all word-pairs. It may also be worthwhile to run at least one word pair where the only trigger is spelling. The spelling strategies used for the measures are tabulated in Table III. The nonsense words themselves are found in Table IV.

Measure	Native endpoint	Foreign endpoint
$\hat{\jmath} < Z < \check{z}$	"j" finally and w/ native trigger: "dge"	"ge"
$æ < A < a$	"aCC"	"ah"
$eyt < E < ey$	w/ native trigger: "ait" w/ foreign trigger: "et"	w/ native trigger: "ay" w/ foreign trigger: "e"
$owt < O < ow$	w/ native trigger: "ote" w/ foreign trigger: "ot"	w/ native trigger: "ow" w/ foreign trigger: "o"

Table III. Spelling of measures.

	Pair	trigger	measure	spelling
i.	*fiZ|lo* (initial stress) *fiZ|ló* (final stress)	Stress only	*ĵ < Z < ž*	"feejelow"~"feegelow" "fijelo"~"figelo"
ii.	*Zov|we* (initial *Zovie* stress)	*-ier*	*ĵ < Z < ž*	"jovaway"~"geovaway" "jauvier"~"geauvier"
iii.	*ĵIb|wE* (initial *žIb|wE* stress)	*ž*	*eyt < E < ey*	"jibaway"~"jibawait" "gibawwe"~"gibawwet"
iv.	*Z̦i̦k̦|l* (initial stress) *Z̦i̦kεl* (final stress)	Stress (*-elle*)	*ĵ < Z < ž*	"joockle" ~"geoockle" "juquelle"~"geuquelle"
v.	*nAb|teyĵ* (initial *nAb|teyž* stress)	*ž*	*æ < A < a*	"nabbatage"~"nahbatage" "nabbetayge"~"nahbetayge"
vi.	*Z̦|bœm|net* *Z̦|bœm|nét*	Stress (*-ette*) & *a*	*ĵ < Z < ž*	"jebamunnit"~"gebamunnit" "jebamenatte"~"gebamanette"
vii.	*Z̦|bám|net* *Z̦|bam|nét*			"jebahmunnit"~"gebahmunnit" "jebahmanette"~"gebahmanette"
viii.	*mAr|lék* *mAr|lét*	*-ette* & Spelling	*æ < A < a*	"marraleck"~"mahraleck" "marralett"~"mahralett"
ix.				"marraleque"~"mahraleque" "marralette"~"mahralette"
x.	*kAst|rben* (initial stress)	Spelling only	*æ < A < a*	"kassterben" ~"kahsterben" "quassteaurbenne" ~"quahsteaurbenne"
xi.	*sobulE* (final stress)	Spelling only	*eyt < E < ey*	"soughbulay"~"soughbulait" "ceauboule"~"ceauboulet"
xii.	*tæĵ|lO* (initial *taĵ|lO* stress)	*a & ž*	*owt < O < ow*	"tadgelow"~"tadgelote" "tahjelo"~"tahjelot"
xiii.	*tæž|lO* *taž|lO*			"taggelow"~"taggelote" "tahgelo"~"tahgelot"

Table IV. Materials.

A final issue is distance and direction between trigger and measure. Pitt and McQueen (1998) have documented effects where the perception of a segment is biased by a preceding segment based on the conditional probability in the lexicon of one segment following following the other. Such perceptual bias effects are known as transitional probability (TP) effects. Moreton (2000) tested theories of TP effects using environments of one or three segments away. The results of Moreton and Amano (1999) suggest that stratal bias effects can occur at distances of four or five segments. As Moreton (2000: p. 68) points out, as the scope of the environment included in the definition of "transitional" grows, TP effects and lexical effects become indistinguishable. Of course, the stratal effects can not be lexical effects in the strict sense, since the words used are not lexical items. They perhaps are some kind of word-to-word analogy being extended from actual lexical items to nonsense strings that contain some of the same segments. The point is that this kind of analogy can be extended from lexical items to nonce words that share certain phonotactic properties in any position in the word.

Even in Moreton and Amano (1999), however, the cues always preceded the measures. If such stratal bias effects could be shown in nonce words where the measure preceded the cue, this would further undermine the account under which stratal bias is a derived effect of transitional probability. It would suggest that on the basis of the unambiguous presence of the phonotactic cue, the word is classified as a whole as belonging to a certain lexical stratum.

The current experiment uses a range of distances between triggers and measures within the word, with the triggers both following and preceding the measures. The experiment uses mostly three-syllable words of various prosodic shapes, with a few two-syllable words to test the effect of word size. Neither of these factors is fully crossed. Moreton (2000) has suggested on the basis of several experiments that grammatical effects are of a larger size than probabilistic effects like transitional probability and lexical effects. If length and distance prove not to be significant sources of variation, such a result would underscore the robustness and of the stratal bias effect as a non-derived grammatical effect.

2.2. Predictions

The predicted effect is that in the foreign condition of each pair (low back *a*, presence of *ž*, French suffixes, final stress, absence of final *t*, "foreign" spelling, or any combination of these cues), the perceptual boundary of the measure will be shifted towards the foreign endpoint (*ž*, low back *a*, absence of final *t*). In the case of combinations of cues, the perceptual shift is predicted to be additive.

2.3. Method

The method is similar to that of Moreton and Amano (1999). Since English orthography does not clearly and consistently represent the difference between front *æ* and back *a* and between the affricate *j* and the fricative *ž*, a brief lecture was included to familiarize participants with the intended sound-spelling correspondences and a pre-training block preceded the training block.

2.3.1. Stimuli

All stimuli, including those used in the pre-training block, were pronounced by a trained phonetician who is a native speaker of English and were recorded in a sound-attenuated booth. The stimuli for the pre-training block were digitized at 44.1kHz and not resampled or otherwise modified. All other stimuli were later resampled to 16kHz.

The *t~ø* continua were made in the following way. Two similar tokens were chosen: one ending in naturally produced *t* and one ending in an open vowel. From the *t*-final utterance, the last four periods of the vowel followed by the release burst were cut. The periods of the vowel were faded slightly. Then, from the vowel-final utterance, the final trailing part of the vowel was removed and the faded vowel periods and release burst from the *t*-final utterance were spliced in. This spliced token served as the most *t*-like endpoint of the continuum, although it was not the same as the naturally produced *t*-final utterance. To create the other members of the continuum, the release burst (but not the preceding vowel periods) was repeatedly faded 50%. Thus the second member of the continuum had a 50% release burst, the third member 25%, etc. For the sixth and most open-vowel-like member of the continuum, the release burst was completely replaced by silence. Thus, it was not the same as the naturally produced open-vowel utterance.

The *ǰ~ž* continua where the *ǰ~ž* segment was word-initial were made in the following way. A single token of a naturally produced utterance containing *ž* was chosen. Three voicing periods of voiced frication were repeatedly removed from the naturally produced *ž*. With most of the voicing removed, the shortest member of the continuum sounded more or less like word-initial *ǰ*.

The *ǰ~ž* continua where the *ǰ~ž* segment was word-medial were made in a slightly different way. A single token of a naturally produced utterance containing *ž* was chosen. Three voicing periods of voiced frication were repeatedly removed from the naturally produced initial *ž*. However, for each three periods of fricative voicing removed, three periods of voicing closure were added. The voicing closure came from a naturally produced utterance containing *ǰ*.

Finally, the *æ~a* continua were made in the following way. Similar natural utterances of both vowels were selected. The portion of the vowel, including formant transitions on both sides and even adjacent consonants in full if they were sonorants, was removed. LPC functions were extracted from these two vowel portions. These two LPC functions were used to compute four intermediate LPC functions. The resulting six LPC functions were used as filters to resynthesize six new vowel portions from a sound source extracted from the original vowel. The resulting six new vowels sounded more or less like a six-step continuum between *æ* and *a*, including the relevant formant transitions. The remaining portion of the word was then pasted back on to the resynthesized portion.

All cutting and pasting and fading operations were done with CoolEditPro software on a PC. The vowel resynthesis was done using Praat (Boersma and Weenink 1992). It was found that Praat worked best with 16kHz sampling. All materials used in the testing blocks were then downsampled to 16kHz so that stimuli within the testing block would not sound incongruent.

2.3.2. Procedure

Participants first heard a brief lecture explaining the differences between *æ* and *a* and between *ǰ* and *ž*. They were then prompted to give their own pronunciations of the spellings "lote/low, nabb/nahb, jove/geove," which were corrected if they did not correspond to the phonetic forms *lot/lo, næb/nab, jov/žov*. Participants were instructed that in all stages of the experiment, they would be asked to judge only between the spellings "t/ø, a/ah, j/ge" which would always have the same values as in the words they had just been asked to pronounce. They were told that the words they would be hearing would be longer than these examples and would be accompanied by two possible spellings which would differ only along the lines of one of these three two-way contrasts.

They were told that the differing portions in each pair of spellings would always be highlighted and that the non-highlighted letters would always be identical in each pair and could be ignored.

The pre-training block consisted of single syllable words only. These syllables were in fact subsets of the longer words used in the other blocks. However, the sound stimuli used in the pre-training block were not the same as the sound stimuli used in the other blocks. Instead, they were naturally recorded tokens of the single syllables (*low, nabb, jove*, etc.). For either member of a pair of syllables (*lot/lo*, etc.), the listener was presented with the same two spelling options. The more native of the spellings (*lot*, etc.) was always on the left side of the screen and the more foreign of the two (*lo*, etc.) was always on the right. The differing portions of each spelling pair were highlighted in yellow. After the listener responded by pressing one of two buttons on a button box, the highlighted portion of the correct spelling turned green and the highlighted portion of the incorrect spelling turned red. Simultaneously, if the response had been correct, a smiley face icon appeared between the two words. If the response had been incorrect, an unhappy face icon appeared instead. As in all blocks, the maximum response time for each trial was two seconds and the interstimulus interval was 0.75 seconds.

The pre-training block was followed by a training block that used all of the actual endpoint stimuli from the testing blocks. It did not contain any intermediate stimuli and it gave the listener the same kind of feedback as in the pre-training block. In both the pre-training and training blocks, each trial was repeated until a correct response was recorded.

The testing block contained one token of each point of each continua. Since there was no feedback, it was harder for listeners to know when or if their response was recorded. To help minimize the number of non-responses, a question mark appeared in the middle of a blank screen following the trial in the case that the listener did not record a response in the necessary two-second window. Each testing block was subdivided into three sub-blocks according to the three measure-types (*t~ø, a~ah, j~ge*). Participants could take a self-timed break between each two sub-blocks. After the first testing block, listeners sat through a series of four back-to-back testing blocks with only self-timed breaks between them. Each listener sat through a total of five testing blocks.

2.3.3. Participants

22 listeners were recruited from among students taking Linguistics courses at U-Mass, Amherst. 12 were women and 10 were men. All were native speakers of English. One reported hearing problems before the age of 10 but not since. Participants received course credit for their participation.

2.4. Results

2.4.1. Overview

First of all, one-tailed t-tests were run for the thirteen word pairs as listed in table IV. For each listener, the total percentage of tokens identified as the foreign endpoint of the measure was computed for both the foreign and native conditions. The percentage

Ben Gelbart

computed from the native condition was then subtracted from the percentage computed for the foreign condition. This percentage then represents the percentage points by which that listener was more likely to identify the measure as foreign if the trigger was foreign. In the case that the difference was negative, no test was run since predictions had not been met. These t-tests were not adjusted for familywise error since they were all pre-planned.

	Pair	Mean % difference	s.d.	p(t) (1-tailed)
i.	fiZ¦lo (initial stress) fiZ¦ló (final stress)	9.090909	10.55005	0.000294**
ii.	Zov¦we Zovie	3.181818	15.95222	0.180071
iii.	ĵ*i*b¦wE ž*í*¦b¦wE	4.44E-16	7.628962	0.5
iv.	Z$_j$k¦l (initial stress) Z$_j$kɛl (final stress)	11.81818	13.82916	0.000318**
v.	nAb¦teyĵ nAb¦teyž	2.424242	8.556183	0.099063*
vi.	Z¦bæm¦net Z¦bæm¦nét (fin. str.)	10.18182	11.09112	0.000156**
vii.	Z¦bám¦net Z¦bam¦nét (fin. str.)	8.181818	11.61131	0.001685**
viii.	mAr¦lék (native mAr¦lét spelling)	2.727273	7.533794	0.052144*
ix.	mAr¦lék (foreign mAr¦lét spelling)	3.181818	9.892398	0.073145*
x.	kAst¦rben (spelling only)	-4.54545	8.640097	-
xi.	sobulE (spelling only)	-0.45455	9.556042	-
xii.	tæj¦lO taj¦lO	2.575758	10.2318	0.125448
xiii.	tæž¦lO taž¦lO	3.636364	8.783648	0.032849**

Table V. Mean differences between percent identifications. For each word pair, the native member is given first. The capitalized letter represents the measure. "Mean % difference" is the difference between the mean percent of the measure identified as foreign in the foreign condition and the mean percent of this segment identified as foreign in the native condition. (**) represents significance at the 5% CL. (*) represents significance at the 10% CL.

The difference percentages are largely positive as predicted. Only two out of the thirteen are negative (x,xi) and these are the only two pairs that differed only in spelling. One pair (iii) had no difference percentage. The remaining ten pairs all had clearly positive shifts. Five of these reached significance at the 5% confidence interval. Of these five, the four that used stress as a trigger had effect sizes much larger than those of any of the other pairs. They also all used the *j/ž* continuum as a measure. Another three pairs reached significance at the 10% confidence level only. These three pairs were the only pairs that used the *a/æ* continuum as a measure. The only pair that used the *t/ø* continuum as a measure and reached significance at the 5% CL was (xiii), which contained two triggers.

These results suggest consistent perceptual bias effects, but with a clear hierarchy of triggers and measures in terms of their robustness. Stress appears to be the best trigger, and spelling the weakest. Among the measures, *j/ž* seems to be the best, followed by *a/æ*, with *t/ø* being the weakest.

2.4.2. Comparing triggers and measures

Since the triggers and measures are not fully crossed, it was impossible to test these hierarchies directly. Instead, three separate averages were computed over the thirteen pairs, one for each measure type. A repeated measures test was then run on these three averages. This showed the type of measure to be significant: $F(2,42) = 5.466$ ($p < 0.01$). The results are shown graphically in Figure I.

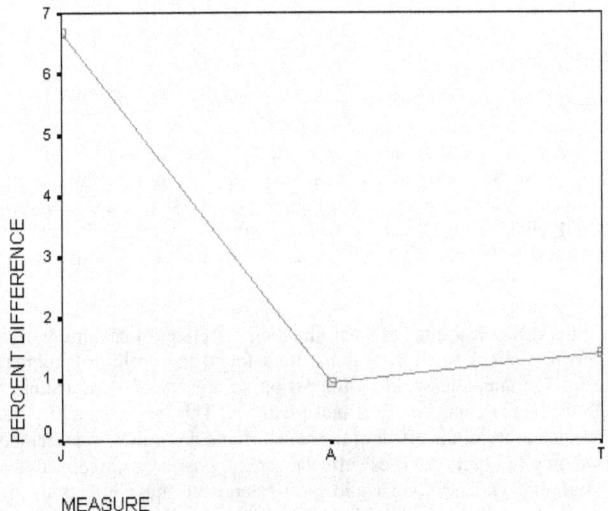

Figure I. Comparison of the three measures. J = *j~ž*; A = *a~æ*; T = *t~ø*.

The effect of measure as tested in this way is highly influenced by the asymmetrical nature of the design. The J-condition as shown in Figure I is certainly greatly inflated since it is the only measure used to measure stress as a trigger. Perhaps

Ben Gelbart

trigger type has a greater effect than measure type. The effect of trigger type, measured in a similar way, did prove more significant: F(4,84) = 20.439 (p < 0.001). Stress indeed did have a far greater impact than any of the other triggers, as shown in figure II.

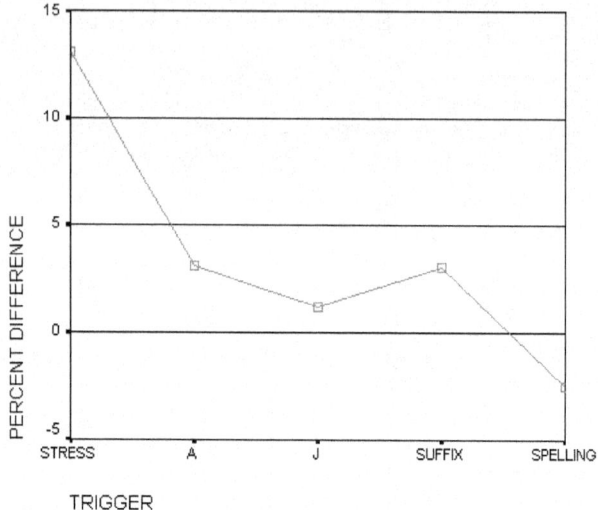

TRIGGER

Figure II. Comparison of triggers. J = *j~ž*; A = *a~æ*.

A few words are due regarding the negative effect of spelling as shown in Figure II and Table V. It is of course absurd to speak of "spelling" having a negative effect. It appears that the particular spelling combinations used in pairs (x-xi) had a negative effect. It is impossible to tell what kind of effect the spelling differences in the other pairs had, since it was confounded with the effects of actual sound differences. What this does suggest is two things.

First of all, it is difficult to predict what direction the effect of spelling will take. The negative effect in pair (x) suggests that listeners found the spelling combination "kass" to be more foreign than "quass." It is impossible to argue with such a subjective judgment, although this is the opposite of the bias predicted. This unpredictability stands in contrast to the predictability of the effect of the actual sound differences represented by other four triggers shown in Figure V. These effects were all positive as predicted. It was thought that the confound with spelling could only exaggerate these effects of sound differences. In retrospect, it seems equally likely that the confound with spelling may be mitigating these effects.

The second point to be made is that the effect of spelling is in any case not great. Whether it hinders or helps the effect of sound differences, it does not do so much. Indeed, all of the effects of the sound triggers and measures were positive despite the presence of spelling confounds in all of them. As shown in Figures I and II, these effects are averaged over several word pairs each, such that the effects of the spelling confounds, sometimes exaggerating and sometimes mitigating, are probably washed out. The averaged effects were all positive (except for the effect of spelling itself as measured separately in Figure II). There is more variation in Table V, where the difference percentages of the word pairs are given individually. This may be due to the individual exaggerating or mitigating effect of the spelling confound in on each word pair. Even so, the overall trend was still positive.

2.4.3. Additivity of triggers

To recall, six of the thirteen word pairs are actually pairs of pairs, although until now we have not made use of this in the analysis. While the overall design of Experiment 3 is not fully crossed, each of these three pairs of pairs constitutes a small-scale crossed design that allows us to measure possible additivity in the effect of the triggers. Two-by-two repeated measures tests were run separately for each of these pairs of pairs.

Pairs (vi-vii) used stress (combined with the suffix *–ette*) and the vowel *a* as triggers and used *j/ž* as a measure. The raw percentages for *ž* identifications in the four conditions were coded and used in a two-by-two repeated measures test. Both main effects were significant: for stress $F(1,21) = 24.143$ ($p < 0.001$) and for A: $F(1,21) = 7.372$ ($p < 0.015$). This again suggests that stress was the more robust trigger. The interaction was not significant: $F(1,21) = 0.422$. This is highly suggestive of an additive effect. The results are shown graphically in figure III.

Ben Gelbart

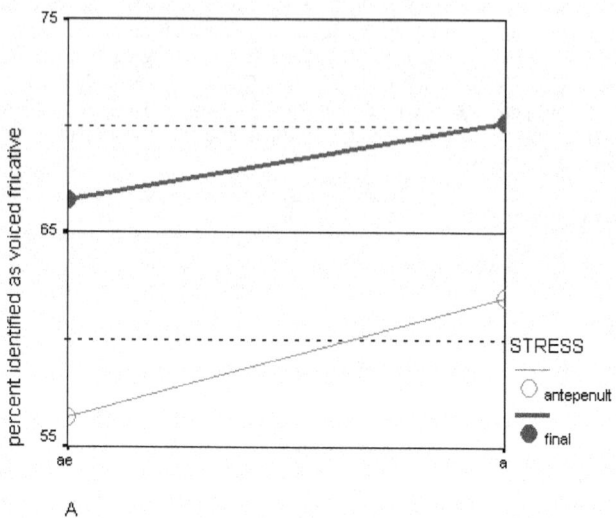

A

Figure III. Percent identified foreign in pairs (vi-vii).

Pairs (viii-ix) used the suffix *–ette* (independently of stress) and spelling as triggers and used *æ/a* as a measure. The raw percentages for *a* identifications in the four conditions were coded and used in a two-by-two repeated measures test. The main effect of the suffix was marginally significant $F(1,21) = 4.272$ (p < 0.06). The main effect of spelling was not $F(1,21) = 0.069$. This again suggests that spelling was not a robust trigger. The interaction was not significant: $F(1,21) = 0.035$. The results are shown graphically in figure IV.

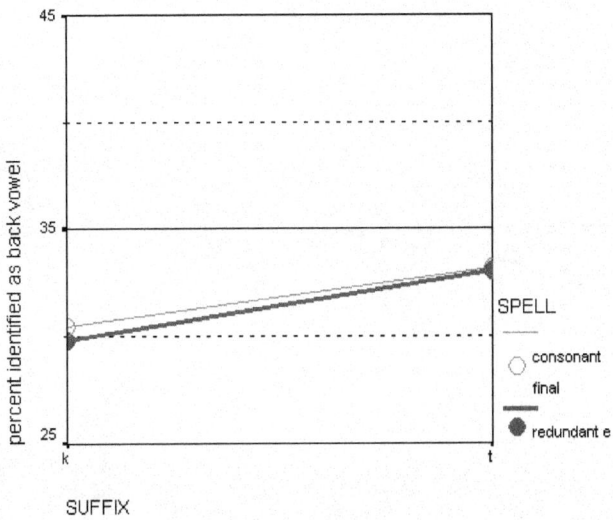

Figure IV. Percent identified foreign in pairs (viii-ix).

Pairs (xii-xiii) used the voiced fricative *ž* and the back vowel *a* as triggers and used *t/ø* as a measure. The raw percentages for *t* identifications in the four conditions were coded and used in a two-by-two repeated measures test. The main effect of J was significant $F(1,21) = 10.502$ (p < 0.005). The main effect of A was marginally significant $F(1,21) = 3.986$ (p < 0.06). This suggests that J is a better trigger than A. This comparison could not be made accurately until now, when all else can be held constant. The interaction was not significant: $F(1,21) = 0.164$. This again suggests additivity of triggers. The results are shown graphically in figure V.

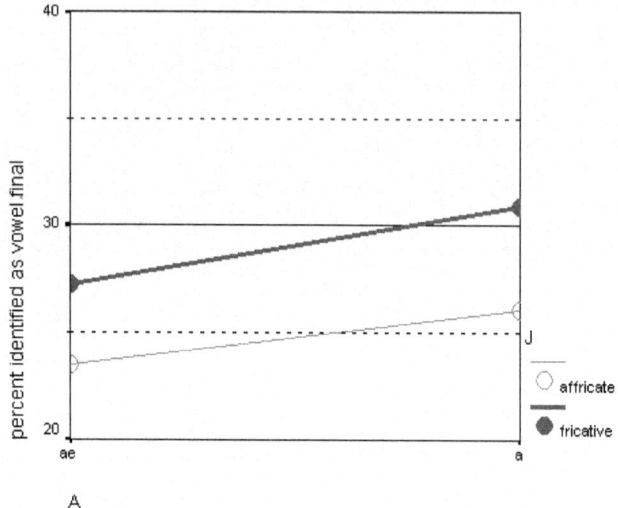

Figure V. Percent identified foreign in pairs (xii-xiii).

Eyeballing figures III-V suggests that the effect of stress as a trigger is about twice that of *a* or *ž* or the suffix *–ette*, while the effect of spelling is negligible. Where two triggers both have an effect, these effects seem additive.

2.4.4. Issues of direction and distance

The word pairs contained a variety of distances between triggers and measures. Furthermore, sometimes the trigger followed and sometimes it preceded the measure. The design was not crossed in such a way as to allow these effects to be tested statistically. However, an eyeball analysis suggests some answers. Information from Tables IV and V is repeated as Table VI for convenience, with the word pairs ordered for distance size. Pairs (x-xi) have been removed, since they used spelling as the only cue, and no distance can be computed. This analysis is highly speculative and can't adjust for the asymmetric confounds of distance and direction with the various triggers and measures.

	Pair	Distance	p(t) (1-tailed)
iii.	*ʃlb⌐wE* *žlb⌐wE*	5	0.5
xii.	*tæj⌐lO* *taj⌐lO*	4	0.125448
xiii.	*tæž⌐lO* *taž⌐lO*	4/3	0.032849**
i.	*ʃiZ⌐lo* *ʃiZ⌐ló*	-1/3	0.000294**
iv.	*Zⱼk⌐l* *Zⱼkɛl*	-1/-3	0.000318**
ii.	*Zov⌐we* *Zovie*	-3	0.180071
vi.	*Z⌐bæm⌐net* *Z⌐bæm⌐nét*	-3/-7	0.000156**
vii.	*Z⌐bám⌐net* *Z⌐bam⌐nét*	-3/-7	0.001685**
viii.	*mAr⌐lék* *mAr⌐lét*	-5	0.052144*
ix.	*mAr⌐lék* *mAr⌐lét*	-5	0.073145*
v.	*nAb⌐teyĵ* *nAb⌐teyž*	-6	0.099063*

Table VI. Distance magnitude and polarity, followed by one-tailed significance for the matched t-score. (**) represents significance at the 5% CL. (*) represents significance at the 10% CL.

Only three pairs had positive distance (trigger preceding measure). Of these, only one reached significance. This is certainly a worse ratio than for the pairs with negative distance. This is certainly surprising, since Moreton and Amano (1999) used only positive distance and obtained largely significant results. The current trend, however, is highly confounded. All of the pairs that used stress as a trigger had negative distance (although pair (i) is somewhat hard to classify). Stress has already been shown to be the most robust trigger. Also, the three pairs with the positive distance were also the only three pairs that used final *t* as a measure. It is possible that this was a weaker measure than the other two. In any case, there is no theoretical reason to believe that positive distance would mitigate the effect. There had been some concern negative distance may mitigate the effect, since listeners may make a judgment before hearing the trigger. At the very least, we can say that this has definitely not been the case.

By comparison, distance might have played some role. Excluding pairs that used stress as a trigger, since this makes distance hard to classify, the pairs that had distances of five or greater were at best only marginally significant. On the other hand, pairs (ii) and (xii) had distances of less than five and were not significant. It is possible that there is some correlation, but this is impossible to measure any more accurately due to the many confounds. If there is an effect of distance, it is in any case gradual, not categorical.

2.5. Discussion

The experiment provides robust evidence for a perceptual bias along lexical strata. The evidence suggests that the stratal bias effect is distinct from either orthographical or transitional probability effects. While all word pairs had orthographical differences, some pairs allowed orthography to be tested without sound differences. Two pairs had only orthographical differences, and these pairs were the only ones to elicit a negative effect. In the three pairs of pairs designs, the only main effect that was not significant was the orthographic effect.

It is possible that the effect is mitigated when the trigger and measure are distant, but this does not appear to be categorical. Possibly, the stratal effect is compounded with a TP effect when the distance between the trigger and measure is small. But even pairs with a distance of five segments had a marginally significant effect. These may be long distance TP effects, which is to say lexical or word-to-word analogy effects. What would be harder to explain in terms of transitional probability is that most of the current results were obtained when the measure actually preceded the trigger. This suggests not transitional probability, but a conditional probability with scope over the entire word. In other words, a categorical distinction at the word level between foreign and native items.

In addition to general support for the stratal bias effect, the experiment also provides evidence of some of the particular mechanics of this effect. First of all, it is clear that not all cues to stratal affiliation make an equal contribution to the perceptual bias. In particular, final stress is clearly a cue of greater magnitude than segmental cues. This is true even though a large number of native words do have final stress in English.

The Celex (Baayen, Piepenbrock and Gulikers 1995) English Lemma database was used to estimate the frequency of these cues in the lexicon. The database contained a total of 52447 entries. Of these, 7422 were found to have final stress in a least one listed pronunciation. This excludes monosyllabic words. Thus, the probability of final stress is roughly 14%. There were 573 instances of $\check{z}$, making its probability about 1%. There were 3668 instances of the low back vowel a, about 7%. Yet a and $\check{z}$ were both relatively weak cues, while final stress was always associated with robust effects.

This would suggest that there is more to stratal cues than mere computation of statistics from the lexicon. Yet it is not clear if a grammatical account can model this difference in effect size any better. Moreton (2000) accounted for the difference in effect size of the initial clusters dl and bw in English by arguing that the former was grammatically banned from English, while the latter was an accidental gap. Thus, the large effect of dl was a grammatical effect, while the less robust effect of bw was merely a transitional probability effect. This explanation cannot be used to account for the differences in effect size between final stress and $\check{z}$ as stratal cues. Neither of these phenomena has zero frequency in English therefore neither can be grammatically banned.

The difference in effect size is apparently the result of the greater prominence of supersegmental cues over segmental ones. All else being equal, it seems that a supersegmental cue to foreignness is more likely to cause a novel word to be categorized as foreign than a segmental cue. In other words, robustness of cues may have more to do with their perceptual robustness than with their grammatical or statistical robustness.

Another result suggests that cues to stratal affiliation have an additive effect. This replicates findings of Moreton and Amano (1999). That similar results were found in English and Japanese by different experimenters is especially suggestive. Moreton (2000: 238) suggests that this does not necessarily mean that classification of words into lexical strata is not categorical. It may be that each instance of classification is categorical, but that the probability of such a classification being made on any given trial is greater when more cues are present.

One question that has not been asked so far is why certain phonotactic features are considered foreign in the first place. Moreton and Amano (1999) have already pointed out that their results are problematic for grammatical theories that assume that a cue must be restricted to a particular stratum. They obtained perceptual shifts away from the Foreign stratum by using the Sino-Japanese cues r^y and h^y. In fact, these cues are not banned from the Foreign stratum in Japanese, they are just far more common in the Sino-Japanese stratum. This is true for the English cues used in the current experiment as well. While the strata in English may not be as clearly delineated as those of Japanese, no one would say that either $æ$ or a is restricted from either native or foreign words in English. And yet this contrast was associated with a perceptual shift of several different measures. Either one or both of these segments is being treated by English speakers as a stratal cue.

A probablistic approach would assume that two features might become correlated with the same stratum because they tend to occur together in the lexicon. Is this true of the cues we have seen to be effective?

One of the most robust effects observed in the experiment was the effect of final stress on the perception of $ž$. Using the Celex English Lemma database again, the probability of $ž$ given final stress is 0.0128. The probability of $ž$ given any other kind of stress is 0.010616, slightly less.

Thus, it seems that the statistical relation is indeed a good predictor of stratal correlation. However, it is pretty easy to see that such small differences in conditional probability are very common in the lexicon if all pairs of potential cues are considered. For instance, if we compute the conditional probabilities of $č$ given final stress in the same way as we did for $ž$, we find that the probability of $č$ given final stress is 0.040825, while the probability of $č$ given any other kind of stress is 0.038867. This is a difference of about the same magnitude as for $ž$.

If conditional probabilities of such small magnitudes are enough to lead to lexical stratification, then perhaps the lexicon is far more stratified than has previously been assumed. In other words, perhaps the *reductio ad absurdum* scenario sketched by skeptics like Inkelas and Zoll (2003) and Rice (1997) is not so absurd after all. This is of course an empirical question. The results of the experiment already suggest that such small differences as whether a stress-final word ends in *t* or *k* are enough to effect whether a vowel earlier in the word is perceived as *æ* or *a* (see Table IV), although none of these four phenomena is particularly rare in any stratum of English.

If however, it turns out that not all statistical imbalances of such a small size are associated with perceptual shifts, we may still be able to salvage the probabilistic account of lexical stratum by limiting how common a cue can be while still being considered a stratal cue. Pater (2004) suggests a limit on the numerical count of words: "if the set of words targeted by a constraint is greater than x, remove indexation." This is intuitive, since we would not want to consider the more common phonotactic cues of the language to be cues to particular strata. Even if these cues are foreign in origin at some date early in the history of the borrowing language, we would assume that once they become very common, they are no longer treated as exceptional by speakers. In fact, the count for ž is only 573 in the lemma database, while the count for č is 2053. Thus, to justify predicting that ž is a stratal cue while č is not, we may set the maximum lemma count for exceptional values at 2000. However, we would probably not want to set it that low. Consider that there are 7422 instances of final stress, and we know for certain that that is a stratal cue. But even if we restrict this limit to segmental cues, we still need to allow a limit of at least 3000, which would include č. Consider the contrast *æ~a*. We never really knew if it was *æ* that was a cue to nativeness, or *a* that was a cue to foreignness, or perhaps both. To be parsimonious, we will only consider as a cue the low vowel with the lowest lemma count. This is *a*, with a count of 3668.

Thus, unless the lexicon is far more stratified than previously considered, even here, the reason that ž and final-stress are perceived as belonging to the same stratum cannot be due solely to statistical computations over the lexicon. Such a correlation may be necessary, as may be the relative scarcity of both properties, but it is not sufficient.

Another problem with trying to reduce lexical stratification to conditional probabilities is the issue of multi-dimensionality. In early attempts to quantify foreignness, Greenberg and Jenkins (1964) imagined foreignness to be a one-dimensional continuum stretching from a native point "1," to higher and higher degrees of foreignness. Ito and Mester (1998) conceive of a quantized series of discrete strata that span an equally one-dimensional line. Nativization patterns of loans from a certain source often proceed in such a fashion, as suggested by Ito and Mester (1998) for Japanese, Holden (1976) for Russian, and Schwarzwald (1998) for Modern Hebrew. However, there may be several distinct sources of loans in a particular language, each with its own unique dimension of such nativization patterns. Thus, to determine the foreignness of a particular form, it is not enough to specify a quantity, but also a direction. Such a scenario emerges from the results of Moreton and Amano (1999), where Sino-Japanese cues and Foreign cues both behave exceptionally, but in different ways. Something similar might be going on in the current experiment, where speakers of English perceived a word as more foreign if it contained *a* instead of *æ*. Since *a* is not banned from native words in English, this effect may be due to English speakers perceiving *æ* as being banned from foreign words.

A purely probabilistic model of the stratified lexicon, even if it could tell which rare and correlated phenomena were stratal cues, still couldn't capture this multidirectionality of exceptional strata. It would not be able to ascertain which strata they were cues to. Consider a cue like θ in English. The lemma count for θ is 1396, lower than any limit we would impose on exceptional cues. Furthermore, we have reason to believe θ may indeed be a stratal cue, but to the native stratum of English. Yet its probability given final stress is 0.028564, which is greater than its probability given non-final stress: 0.026297. The same is true of the syllabic retroflex American English $\tilde{a}$, a native cue if there ever was one. It has a count of 2509. Its probability given final stress is 0.05605. Its probability given non-final stress is lower: 0.046485. On the other hand, the voiced interdental fricative ∂ does have lower probability given final stress, 0.009162 compared to a whopping 0.009262 given non-final stress, and thus might be correctly classified as a native cue by a purely probabilistic model. The other two presumably native cues: θ and $\tilde{a}$, would be classified as foreign cues, as would other segments like $\check{c}$ that probably are not cues of any kind.

The problem of how phonotactic exceptionalities are recognized as cues is similar to the problem of multi-dimensionality. For instance, how do Japanese speakers decide that p is a Foreign stratum cue, while r^y is a Sino-Japanese cue? Ito and Mester (1998) solve this dilemma by assuming the different strata are merely a matter of degree: weak cues are Sino-Japanese, whereas strong cues are Foreign. Yet Moreton and Amano (1999) have shown that these two kinds of cues are associated with boundary shifts in opposite directions. Studying the surface forms can reveal which phonotactic features are rare, but in order to know which features and which lexical items are associated with which strata, we also need to know what kinds of strata speakers suppose to exist in the first place.

The similarities between lexical strata and statistical probabilities in the lexicon are most certainly not coincidental. However, they are also clearly not reducible to each other. While lexical strata most certainly have their basis in statistical probabilities in the lexicon, and may dissolve if they lose these probabilities, they seem to be mediated by some non-grammatical, non-computable type of cultural information such as "final stress is a sign of foreignness," "$\check{z}$ is a sign of foreignness," etc. This would simply be some kind of cultural prejudice that can not be deduced from the surface forms of the language alone.

3.   Conclusion

This experiment, like Moreton and Amano (1999), suggests that the stratified phonological lexicon is neither an artifice in the minds of historical linguists nor an aberration in the history of certain languages. It also suggests that the phonological lexicon has a life of its own beyond statistical imbalances that can be deduced from the lexicon.

A relatively common rarity in English, final stress, was shown to be robustly associated with other, more exceptional phenomena in the linguistic behavior of English speakers. These included the low vowels $æ$ (associated with native English words) and a (associated with foreign words) and whether a word was consonant final (associated with native English words) or vowel final (associated with native foreign words), even though none of these properties are banned from words of any particular origin. Furthermore, many of the correlations between these properties were very weak. These findings suggest that while rarity and correlation may be a necessary precondition for exceptionality and stratal affiliation, they may not be sufficient. It also suggests that total segregation of phones between strata may not be necessary for them to be associated with those strata.

The greater robustness of final stress as a cue to foreignness compared to other cues like *ž* suggests that some features are more strongly associated than others with certain strata. Thus, while the relationship between strata may not be hierarchical, each strata may have a hierarchy of features associated with it. This may be the kind of scenario that leads to the "impossible nativizations" described by Ito and Mester (1998). As a thought experiment, consider an analogous case of "impossible hyperforeignization" in English. Janda et al (1994) report some Ohio shoppers pronouncing the "Target" chain of stores as *taržé*. This is total hyperforeignization, as opposed to the received *tárgət*, which represents no hyperforeignization at all. Consider now what would be a more possible example of partial hyperforeignization: *táržət* or *targét*?

An even more pressing issue is how certain phonotactic exceptions come to be cues to particular strata in the first place. If relatively little can be predicted from the surface forms of a language regarding the structure of its stratified lexicon, then how is it that it was possible to predict the observed stratal bias effects? Statistical counts and correlations of phonotactic cues were not consulted in preparation for this experiment. Yet, even after the fact, these statistical distributions were not always able to account for the already observed experimental results. What where consulted beforehand were native speaker intuitions about what sounds and words they considered foreign or native. It was these intuitions which were tested among other native speakers who were naive to the purpose of the experiment. Such intuitions were bolstered by cases of hyperforeignization such as those reported by Janda et al. (1994). It is cases like this, where native speakers reveal a conviction that some sound is more appropriate than another in a word of a certain origin, that make for accurate predictions of stratal bias.

References

Baayen, R., R. Piepenbrock and L. Gulikers. 1995. *The CELEX lexical database* (CD-ROM). Philadelphia: Linguistic Data Consortium.
Becker, M. 2003. *Lexical stratification of Hebrew – the disyllabic maximum*. Proceedings of Israel Association for Theoretical Linguistics 19, edited by Yehuda Falk.
Boersma, P. and D. Weenink. 1992. *Praat: Doing phonetics by computer*. [http://www.fon.hum.uva.nl/praat/]
Greenberg, J. and J. Jenkins. 1964. Studies in the Psychological Correlates of the Sound System of American English. *Word* 20: 157-177
Holden, K. 1976. Assimilation Rates of Borrowings and Phonological Productivity. *Language* 52: 131-147.
Inkelas, S., Orgun, O. and Zoll C. 1997. The Implications of Lexical Exceptions for the Nature of Grammar. In *Derivations and Constraints in Phonology* edited by Iggy Roca. Oxford University Press.

Inkelas, S. and Zoll C. 2003. *Is Grammar Dependence Real?* Ms., University of California, Berkeley and MIT. [http://roa.rutgers.edu/]

Ito, J. and A. Mester. 1998. The Phonological Lexicon. In *The Handbook of Japanese Linguistics* ed. N. Tsujimura. Blackwell.

Janda, R., B. Joseph and N. Jacobs. 1994. Systematic Hyperforeignisms as Maximally External Evidence for Linguistics Rules. In *The Reality of Linguistics Rules.* Amsterdam: John Benjamins.

Kelly, M. 1988a. Rhythmic alternation and lexical stress differences in English. *Cognition* 30: 107-137.

Kelly, M. 1988b. Phonological biases in grammatical category shifts. *Journal of Memory and Language* 27: 343-358.

Moreton, E. 2000. *Phonological Grammar in Speech Perception.* Ph.D. dissertation, University of Massachusetts, Amherst.

Moreton, E. and S. Amano. 1999. *The Effect of Lexical Stratum Phonotactics on the Perception of Japanese Vowel Length.* Ms. University of Massachusetts and NTT Communication Science Laboratories.

Pater, J. 2004. *Learning a Stratified Grammar.* handout at Boston University.

Pitt, M. and J. McQueen. 1998. Is compensation for coarticulation mediated by the lexicon? *Journal of Memory and Language* 39: 347-370.

Rice, Keren. 1997. Japanese NC Clusters and the Redundancy of Postnasal Voicing. *Linguistic Inquiry* 28: 541-551.

Saciuk, B. 1969. The stratal division of the lexicon. *Papers in Linguistics* 1: 464-532.

Schwarzwald, O. 1998. Word Foreignness in Modern Hebrew. *Hebrew Studies* 39: 115-142.

Venezky, R. 1970. *The Structure of English Orthography.* The Hague: Mouton.

Department of Linguistics
South College
University of Massachusetts, Amherst
Amherst, MA 01003

gelbart@linguist.umass.edu

Voicing and geminacy in Japanese: An acoustic and perceptual study[*]

Shigeto Kawahara

University of Massachusetts, Amherst

1.   Introduction

Maintaining voicing in obstruents is articulatorily challenging. During obstruent closure, intraoral air pressure goes up quickly, and as a consequence it becomes difficult to maintain a sufficient transglottal air pressure drop to produce voicing. This difficulty becomes more problematic in geminates, which have long closures (Hayes and Steriade 2004; Jaeger 1978; Ohala 1983; Westbury 1979). Reflecting this articulatory difficulty, historically, Japanese allowed no voiced geminates. Various alternations support this distributional restriction. Coda nasalization in mimetic gemination in (1b) is induced to avoid voiced geminates (Kuroda 1965; Itô and Mester 1999). A root-final vowel in Sino-Japanese is syncopated in compounds, with subsequent place assimilation of the root-final consonant to the following consonant, as shown in (2a) (Itô and Mester 1996); however, such syncope is blocked when it would result in a voiced geminate, as in (2b):

(1)   a. /tapu+μ+ri/ →   [tappuri]   *[tampuri]   'a lot of'
       b. /zabu+μ+ri/ →   [zamburi]   *[zabburi]   'splashing sound'

(2)   a. /hatu+kaku/ →   [hakkatsu]   *[hatsutatsu]   'revelation'
       b. /hatu+gen/ →   [hatsugen]   *[haggen]   'remarks'

Although a restriction against voiced geminates in Japanese is clearly motivated, in recent loanwords, we do find voiced geminates (McCawley 1968; Itô and Mester 1999 among others). A word-final sound in a donor language is geminated, and the following vowel is epenthesized, even when such gemination results in a voiced geminate, as in

---

[*] I am thankful to José Benki, Kathryn Flack, Ben Gelbart, Joe Pater, Chris Potts, John McCarthy, Caren Rotello and Taka Shinya for their comments and suggestions on this project. I would like to thank especially John Kingston for his support in virtually every aspect of the studies reported here. Finally, I wish to express my gratitude for all the Japanese speakers who participated in the experiments reported below. All remaining errors are mine.

Kathryn Flack and Shigeto Kawahara (eds.), UMOP 31, 87-120.

'dog' borrowed as [doggu] (Lovins 1973; Katayama 1998; Shirai 1999; Takagi and Mann 1994). Such a voiced geminate minimally contrasts with a voiceless geminate; there are minimal pairs like [kiddo] 'kid' and [kitto] 'kit' that show that voicing is indeed phonemic for geminates in the loanword phonology. Yet, as mentioned above, voicing in geminates is aerodynamically challenging, and there is evidence that voicing in singletons and voicing in geminates behave differently in Japanese phonology; Nishimura (2003) and Kawahara (2005) show that only voiced geminates, but not voiced singletons, devoice when they cooccur with another voiced obstruent. In other words, only voicing in geminates can be lost in response to the OCP(voi), which prohibits more than one voiced obstruent within a stem (e.g. Itô and Mester 2003).

(3)   Voicing in geminates can optionally be lost in response to the OCP(voi)

gebberusu	~	gepperusu	'Göbbels (proper name)'
guddo	~	gutto	'good'
beddo	~	betto	'bed'
doggu	~	dokku	'dog'
baggu	~	bakku	'bag'

(4)   Voicing in singletons is not lost

bagii	'buggy'	bogii	'bogey'
bobu	'Bob'	doguma	'dogma'
dagu	'Doug'	daibu	'dive'
giga	'giga- (10^9)'	gaburieru	'Gabriel'

To account for this asymmetry, following the P-Map hypothesis (Steriade 2001), Kawahara (2005) hypothesizes that voicing in geminates is more easily lost because voicing in geminates is harder to hear. Cross-linguistically, contrasts that are signaled by weaker cues are more prone to phonological neutralization (Hura et al. 1992; Jun 2004; Kohler 1990). For example, preconsonantal consonants have many disadvantages in signaling their place: they suffer from the lack of CV transitions which provide primary cues for place distinction, and they are often unreleased, which again weakens place cues (see Jun 2004 and references cited therein). As is well-known, preconsonantal consonants undergo place neutralizations much more often than prevocalic consonants.

Kawahara (2005) applies the same logic to explain the contrast between (3) and (4). He hypothesizes that voicing is harder to detect in geminates, and therefore it is more prone to phonological neutralization. More concretely, for example, the [atta]~[adda] contrast is less reliably perceived than the [ata]~[ada] contrast; so it would not have a large perceptible impact if [adda] became [atta], while if [ada] became [ata], it would be more perceptually conspicuous. In other words, neutralizing voicing in geminates is regarded as "perceptually tolerated articulatory simplification" (Kohler 1990): since voicing in geminates is hard to perceive, its loss does not have a large perceptible consequence, hence tolerated. Just as preconsonantal consonants are more likely to undergo place neutralization than prevocalic consonants, perceptually weak voicing in geminates is more easily lost than more robustly cued voicing in singletons. See Kawahara (2005) on why the loss of voicing in (4) cannot be purely due to the articulatory difficulty of voicing in geminates.

This paper reports phonetic studies that aim to verify the hypothesis that voicing in geminates is less reliably perceived than voicing in singletons. As little is known about voicing cues in Japanese voiced geminates, I began with an acoustic experiment that identified a set of acoustic correlates that distinguishes voiceless and voiced consonants. The primary aim of this experiment was to investigate whether such cues manifest themselves differently in singletons and geminates, and if so, in what ways. In other words, the experiment looked for evidence from acoustics bearing on whether voicing is harder to detect in geminates. The result of this experiment shows that some cues are indeed weakened in geminates, which might lead to higher confusability of voicing in geminates.

With these observations in mind, the second experiment more directly tested the core hypothesis of this paper, which is that voicing is harder to detect in geminates than in singletons. In order to replicate most closely the natural environments in which Japanese listeners hear voiced geminates, the natural tokens recorded in the first experiment were used. In the experiment, Japanese speakers identified the presence (or the absence) of voicing in a noisy environment. The result clearly shows that voicing is hard to perceive in geminates, while voicing in singletons is accurately perceived. In summary, the two experiments reported in this paper show the following points:

(5)   a. Some phonetic correlates of voicing are weakened in geminates.
        b. Some phonetic differences are enhanced in vowels next to geminates.
        c. Voicing in geminates is not well perceived.

2.   Experiment I: Acoustics of voicing and geminacy

The first experiment was designed to investigate the following three questions:

(6)   a. What are the phonetic cues that signal voicing in Japanese?
        b. How are such cues different in singleton and geminates?
        c. Do geminates have a disadvantage in signaling voicing?

2.1.   Methods

2.1.1.   Speakers and recording

Three native speakers of Japanese were recruited at the University of Massachusetts, Amherst. They were all female and in their mid twenties. An informed consent form was obtained from each speaker in accordance with the University of Massachusetts human research subjects guidelines. The dialects the subjects spoke were Shizuoka Japanese (Speaker E), Tokyo Japanese (Speaker T) and Hiroshima Japanese (Speaker W). The frame sentence used in the experiment was Standard (Tokyo) Japanese, and the subjects were asked to read the sentences in Standard Japanese as well. They were all paid for their time. The speech was recorded through a microphone (MicroMic II C420 by AKG) by a CD-recorder (TASCAM CD RW-700) at 44.1 KHz sampling rate, in a sound attenuated booth. The recorded tokens were then downsampled to 22.050 KHz and 16 bit quantization level when they were transferred to a pc. Including short breaks between each repetition, the recording session lasted about 45 minutes.

2.1.2. Stimuli

The stimuli consisted of 36 words, which were mostly nonce words.[1] In addition, 36 nonce words were added as fillers. The target words were all disyllabic: the first consonant was [k], the second consonant was the target ([p], [t], [k], [pp], [tt], [kk], [b], [d], [g], [bb], [dd], [gg]) and three different vowels were used ([a], [e], [o]) for both the first syllable and the second syllable (henceforth V1 and V2, respectively); some examples are *kappa, kaba, kege, kokko, kodo*. The speakers were asked to pronounce these tokens with a HL tonal contour, which is a default pattern in loanword and nonce word pronunciation.

Each word was written on a card in *katakana* orthography, which is conventionally used for loanwords. This was because voiced geminates are found only in loanwords. Six repetitions of each set were recorded, with a short break between each repetition. The order of the stimuli was randomized after each repetition. In order to solicit natural utterances and avoid domain-edge strengthening effects on target words (e.g. Fougeron and Keating 1997), the stimuli were embedded in the following frame sentence:

(7)   jyaa ____ de   onegai
        then ____ with   please
        'Please, (do something) with ___. (casual register)'

In order to avoid the hyper-articulation of the materials in an experimental environment, the speakers were encouraged to produce sentences in a natural speech style. Specifically, they were asked to imagine a situation where they were preparing a party and they wanted their friend to fetch the things whose names were the target words.

2.1.3. Measurement and analysis

All measurements were done using Boersma and Weenink's (1992) Praat. Following the past literature on acoustic and perceptual correlates of voicing (Lisker 1987; Kingston and Diehl 1994; Raphael 1981; Stevens and Blumstein 1981), the following values were measured: (i) closure voicing duration; (ii) duration of the preceding vowel; (iii) closure duration; (iv) F0 of the surrounding vowels and (v) F1 of the surrounding vowels, as illustrated in
Figure 1.

Closure voicing is the glottal vibration into the obstruent closure; this acoustically appears as a voice bar, low frequency energy observed during closure near the bottom of a spectrogram. This should appear in only voiced consonants. The second cue lies in the

---

[1] It was impossible to completely exclude real words in this set; [kaka], [kaba], [kakka] are real words. Yet as they were all written in *katakana* orthography, at least [kaka] and [kakka], which are usually written in *hiragana*, should have been hard to recognize as real words.

immediately preceding vowel, which is known to be longer before voiced consonants. The third correlate of voicing is closure duration: cross-linguistically, voiceless consonants are longer than voiced consonants. Finally, F0 and F1 are generally higher next to voiceless consonants in both the preceding and following vowels (V1 and V2, respectively). These measurement points are illustrated in Figure 1:

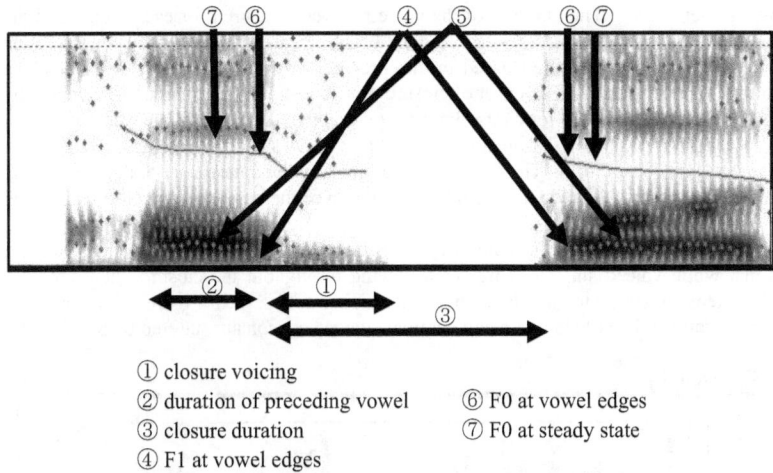

① closure voicing
② duration of preceding vowel   ⑥ F0 at vowel edges
③ closure duration   ⑦ F0 at steady state
④ F1 at vowel edges
⑤ F1 at steady state

Figure 1. Illustration of measurement points. The spectrogram is that of [kobbo] uttered by Speaker E.

More detailed explanations of how these values were measured are provided below.

To analyze these acoustic measures, an ANOVA was run with CONSONANTAL LENGTH[2] (2-level), VOICING (2-level) and SUBJECT (3-level) as independent variables. This is because what is of interest is how a voicing difference manifests itself in these acoustic values, and how they vary in singleton and geminate environments. I treated SUBJECT as an independent variable as well to test for any inter-speaker variability.

2.2. Results

The overall results show that the phonemic difference in voicing is signaled in both singletons and geminates by all of the measurements taken here. However, some of these acoustic correlates, most notably closure voicing, are weakened in geminates. Yet all speakers attempt to compensate for this weakening in some way or another, by enhancing some phonetic differences next to voiced geminates or by producing a phonetic difference not observed in voiced singletons. Each of the acoustic measures will be

---

[2] In this paper, "duration" refers to a phonetic temporal property while "length" refers to a phonological geminacy contrast. SMALL CAPITAL LETTERS are used to represent independent variables.

discussed in more detail below.

2.2.1. Closure voicing

One of the most important voicing cues is the extent to which voicing continues into the closure, acoustically realized as a voice bar (Lisker 1986; Raphael 1981; Stevens and Blumenstein 1981). The duration of the voice bar was measured for each token, and the ratio of closure voicing with respect to duration of closure was calculated. The duration of a voice bar was measured based on the presence of low frequency energy near the bottom of spectrograms. The onset of closure was acoustically unambiguous, signaled by abrupt disappearance of formants. In case of gradual closure, which was sometimes observed for dorsals, the disappearance of F2 and F3 was used as a criterion. The offset was set at the release of the closure, which was cued by the appearance of the burst noise. The values reported here do not include the duration of the burst noise in closure duration.

One of the most noticeable differences between voiced singletons and geminates is that while voiced singletons maintain voicing throughout the closure, there are very few tokens of geminates in which such full voicing is observed. Figure 2 illustrates the representative token of singleton and geminate voiced consonants, uttered by Speaker W:

a. Singleton [b].

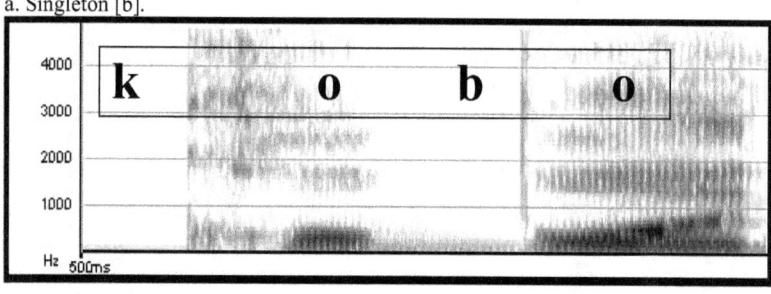

b. Geminate [b]

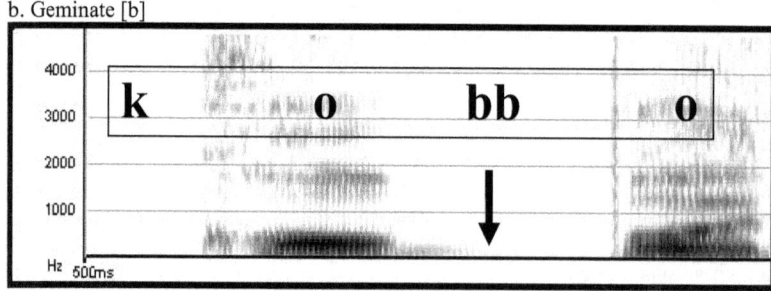

Figure 2. Spectrograms of singleton and geminate [b] pronounced by Speaker W. While voicing is fully maintained in the singleton [b] (a), partial devoicing is observed after the arrow in the geminate [bb] (b).

The first spectrogram is that of a singleton [b], and as seen, closure voicing continues throughout the closure. In contrast, in the second spectrogram of a geminate [bb], voicing stops in an early phase of the closure, at the point indicated by an arrow.

This contrast between singleton and geminate voiced consonants is a very general pattern observed for all speakers; for example, Speaker T shows full voicing for all through 54 singletons, while she exhibits no tokens of geminates in which voicing is maintained more than 80 percent of the closure. Speaker E and Speaker W show two instances of exhaustively voiced geminates, [dd] and [bb], respectively, but all other tokens are partially devoiced.

Consider next Figure 3 which shows the spectrograms of [p] and [pp] pronounced by Speaker W:

a. Singleton [p]

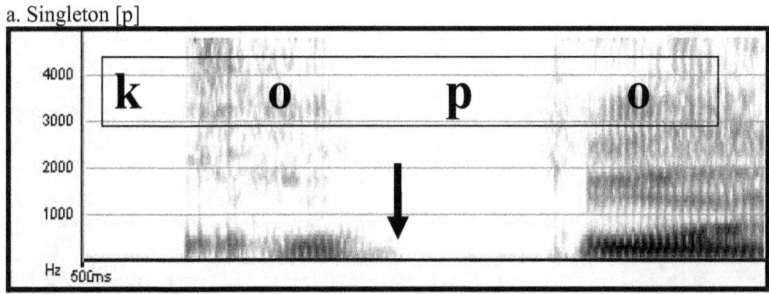

b. Geminate [p]

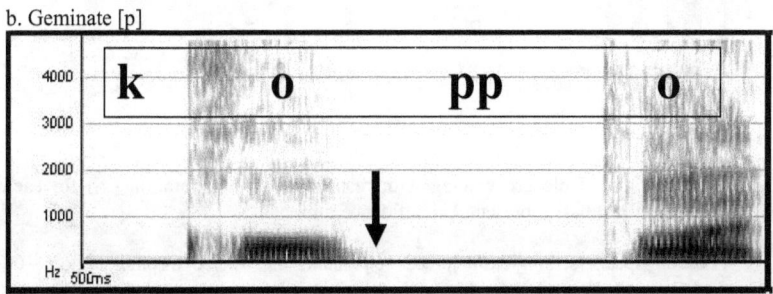

Figure 3. Spectrograms of [p] and [pp] pronounced by Speaker W. Voicing leakage is observed, indicating that voicing cessation and consonantal closure do not completely coincide.

Even voiceless consonants have a small amount of voicing leakage; there is short closure voicing after the closure of a voiceless [p]. For singleton pairs like [p]~[b], even with such voicing leakage in [p], a voicing contrast is still clear, since the closure voicing is exhaustive for [b]. However, in geminate pairs like [pp]~[bb], given that even [pp] has

Shigeto Kawahara

some closure voicing and [bb] is partially devoiced, the acoustic difference between voiced and voiceless consonants is very small.

To analyze these observations numerically, the proportion of closure voicing with respect to closure duration was calculated. The results are summarized in Figure 4. Here and throughout, in summary figures, the first pair of bars in each graph represents singleton values while the second pair shows geminate values. Within each block, the first (solid) bar represents voiced consonants, and the second (striped) bar represents voiceless consonants. Error bars represent 95% confidence intervals, calculated as $t_{0.05} \times$ standard error of the mean (s.e.).

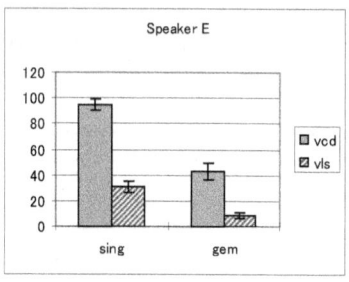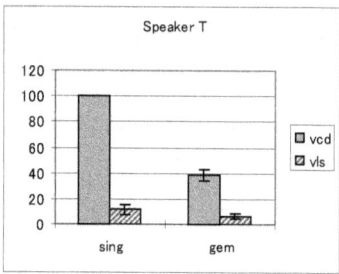

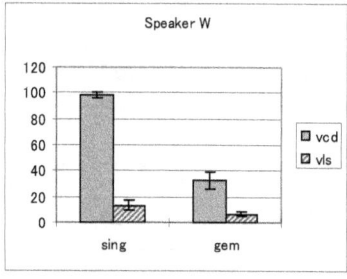

Figure 4. The ratio of closure voicing with respect to closure duration (%) for each speaker. Error bars represent 95 confidence intervals.

For all speakers, singleton voiced consonants are voiced through almost 100 percent of their closure; on the other hand, voiced geminates are voiced through only around 30 to 40 percent of their closure, indicating that partial devoicing is prevalent. The acoustic voicing difference between voiced and voiceless singletons is thus drastically reduced in geminates. Such weakening of closure voicing in geminates should have a strong impact on the perceptibility of voicing in geminates, as closure voicing is presumably an important cue to phonemic voicing (Lisker 1978, 1986; Raphael 1981; Stevens and Blumstein 1981), and about 60 percent of the closure, the geminates are "voiceless." Lisker (1978) has shown that consonants with 120ms closure duration and 40 ms closure voicing, which are quite close to Japanese semi-devoiced geminates, are perceived by English speakers as voiceless about 70 percent of the time, even when other cues such as V1 length (see below) are in favor of "voiced" percept. Further, since such phonemically "voiced" consonants are acoustically "voiceless" at the time of release, this should again attenuate overall voicing perception in geminates, because it is known that the onset cues have primacy over offset cues (e.g. Rapahel 1981; Slis 1986).

The result of ANOVA suggests that VOICING and CONSONANTAL LENGTH both significantly affect the ratio of closure voicing: $F(1, 608)=2073.928$, $p<.0001$ and $F(1, 608)=877.896$, $p<.0001$, respectively. It does not come as a surprise that a phonemic voicing distinction affects the proportion of closure voicing. More interesting is the fact that the LENGTH difference has an effect on closure voicing as well. This is because voiced geminates are frequently partially devoiced, as seen above. This is also indicated by the fact that the LENGTH-VOICE interaction is significant: $F(1, 608)=414.107$, $p<.0001$: only voiced geminates, not voiced singletons, undergo partial devoicing.

Though partial devoicing is prevalent, a voicing contrast seems always maintained: an independent sample t-test shows a significant difference between voiced and voiceless geminates ($t(332)=16.450$, $p<.0001$). Compared to voiced geminates, which have around 30~40 percent closure voicing, voiceless geminates have on average less than 10 percent closure voicing. The average absolute duration of voicing in geminates is around 40 ms across all the speakers (Speaker E=42.2 ms, Speaker T=42.4 ms, Speaker W=38.3 ms), which is small, but not negligible. These values are different from those for voiceless geminates, which are about 10 ms (Speaker E=10.5 ms, Speaker T=9.0 ms, Speaker W=9.7 ms).

2.2.2. Duration of preceding vowels

The second phonetic difference that correlates with the voicing distinction is the duration of the immediately preceding vowel (V1). To measure this, the onset of the V1 is set at the first periodic wave after the aspiration of the preceding [k], judged based on the beginning of a periodic wave in the waveform. The offset is set at the onset of consonantal closure, signaled by the disappearance of F2 and F3.

An ANOVA shows that VOICING and LENGTH both have a statistically significant impact on the duration of V1 ($F(1, 603)=166.344$, $p<.0001$ and $F(1, 603)=453.184$, $p<.0001$). This reflects the tendency for V1 to be longer before voiced consonants as well as before geminates, as illustrated in Figure 5:

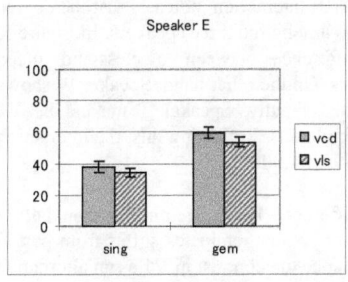 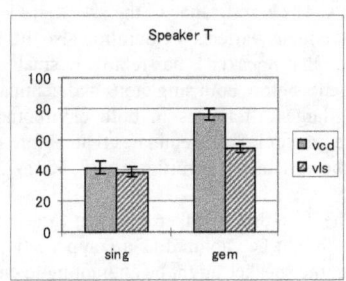

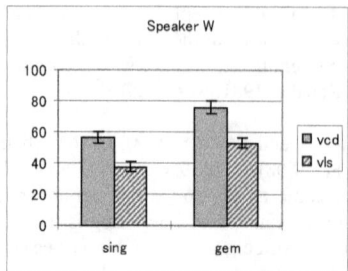

Figure 5. The duration of the preceding vowel (milliseconds).

Vowels are cross-linguistically known to be longer before voiced obstruents than before voiceless obstruents (Chen 1970; Kingston and Diehl 1994; Raphael 1972, 1981). This is true in Japanese for all the speakers both before singletons and geminates, as confirmed by the ANOVA result.

The fact that LENGTH has a main effect on V1 duration captures the tendency for preceding vowels to be longer before geminates. This is contrary to the cross-linguistic tendency that vowels are shorter in closed syllables than in open syllables (see Maddieson 1985; though see Smith 1995 who argues that this tendency is not universal, using data from Japanese). One might suspect that a Japanese geminate does not close a preceding syllable, but this postulation is not tenable because geminates count as moraic, and thus appear to be coda consonants (see e.g. McCawley 1968; Poser 1990 for evidence). Even with such lengthening of V1, however, a voicing contrast is still maintained before geminates: t(322)=11.116, p<.0001.

According to the ANOVA, the interaction of VOICING and LENGTH is significant (F(1, 603)=19.487, p<.0001); this shows that the extent to which voicing affects V1 duration is different for singletons and geminates. This is most clearly observed in Speaker T; the V1 difference due to voicing is larger before geminates. A related observation is the SPEAKER-VOICE interaction and the SPEAKER-LENGTH interaction are both statistically significant (F(2, 603)=22.648, p=.004 and F(2, 603)=4.584, p=.011, respectively). The significance of the SPEAKER-VOICE interaction shows that there is inter-speaker variation for the extent of V1 difference before voiceless versus voiced consonants. The significance of the SPEAKER-LENGTH interaction indicates that the degree to which geminacy affects V1 duration also differs among the three speakers. In Figure 5, we can see that Speaker E has relatively small differences between voiceless and voiced environments before both singletons and geminates. On the other hand, Speaker W shows relatively large differences in both environments. Finally, Speaker T makes the V1 difference greater before geminates than before singletons. Reflecting this, the interaction of all of the variables is significant: F(2, 603)=7.895, p<.0001.

The fact that Speaker T has a larger difference before geminates than before singletons might be captured as a compensation effect: as geminates suffer from partial devoicing, the speaker might be attempting to enhance the contrast in V1 as an alternative means signaling voicing. In other words, to make up for the weakening of closure voicing, she enhances another cue. We observe below that a similar effect is exhibited by the other two speakers in other acoustic dimensions.

2.2.3. Closure duration

The third difference between voiced and voiceless consonants is closure duration. How closure duration was measured is stated in §2.2.1. An ANOVA suggests that, as cross-linguistically often observed (Westbury 1979; Ohala 1983: 195), voiced consonants are shorter in duration than voiceless consonants $(F(1, 602)=182.938, p<.0001)$, which presumably contributes to perception of voicing (Lisker 1957, 1981, 1986; Kingston and Diehl 1994). LENGTH, quite naturally, exhibits a large significance; by definition, geminates have longer closure duration $(F(1, 603)=3220.478, p<.0001)$. The interaction of VOICING and LENGTH is not significant $(F(1, 603)<1)$. This means that the difference in closure duration due to a voicing contrast is about the same in singletons and geminates.

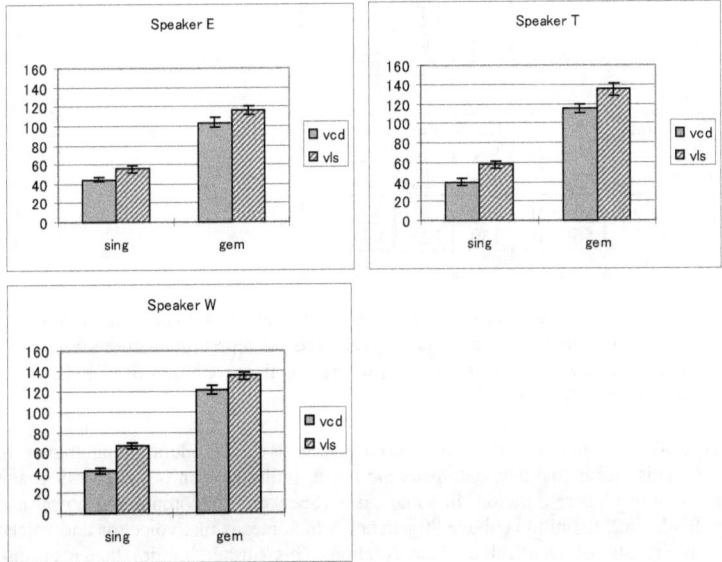

Figure 6. Closure duration of each consonant type (milliseconds).

As revealed by the ANOVA, the closure duration difference is consistently present in singletons and geminates; the absolute magnitude of the differences between voiced and voiceless consonants is about the same in singletons and geminates, as indicated by the fact that the VOICE-LENGTH interaction was non-significant. However, given the consistent difference, geminate pairs are more similar to each other than singleton pairs because geminates have inherently longer duration. To show this numerically, the proportion of voiced consonants' closure duration with respect to voiceless consonants' closure duration was calculated. The following chart summarizes the results at each place of articulation.

a. Speaker E

	Singletons			Geminates		
	Vls (ms)	Vcd (ms)	Vcd/Vls (%)	Vls (ms)	Vcd (ms)	Vcd/Vls (%)
Lab	63	49	78 (13)	115	107	93 (8)
Cor	59	43	72 (15)	120	107	90 (10)
Dor	43	39	92 (9)	113	97	86 (12)

b. Speaker T

	Singletons			Geminates		
	Vls (ms)	Vcd (ms)	Vcd/Vls (%)	Vls (ms)	Vcd (ms)	Vcd/Vls (%)
Lab	66	48	72 (15)	140	118	84 (12)
Cor	52	30	58 (17)	137	107	78 (13)
Dor	55	52	95 (13)	129	119	92 (9)

c. Speaker W

	Singletons			Geminates		
	Vls (ms)	Vcd (ms)	Vcd/Vls (%)	Vls (ms)	Vcd (ms)	Vcd/Vls (%)
Lab	77	50	64 (16)	145	125	86 (16)
Cor	63	35	56 (16)	130	123	94 (15)
Dor	60	39	64 (17)	123	115	94 (14)

Table 1. The ratio of voiced consonant with respect to voiceless consonants in terms of closure duration. The numbers in parentheses represent margins of error, calculated as $t_{0.05}(n-1) \times ((p(1-p)/n)^{0.5})$ where p is the proportion of vcd/vls, and n is the number of data points.

What is evident is that the ratio of voiced/voiceless closure duration is higher in geminates. This means that geminate pairs are more similar to each other than singleton pairs in terms of closure duration. In some cases (Speaker W's coronal and dorsal and Speaker E's labial), the ratio is above 90 percent, which means that voiceless and voiced consonants are almost identical in their duration. This further implies that a closure duration difference, which is presumably one of the perceptual cues for voicing, is harder to detect in geminates. This is yet another factor that might make a voicing distinction in geminates harder to perceive.

2.2.4. F0 at V2 onset

The fourth voicing cue is F0 frequency at the onset of the following vowel (V2). F0 was measured at the first periodic wave right after the consonantal burst, using autocorrelation function of Praat. Cross-linguistically, it is observed that F0 is higher in vowels next to voiceless consonants (see Kingston and Diehl 1994 among others), and this is in general true for the Japanese speakers as well. Figure 7 illustrates:

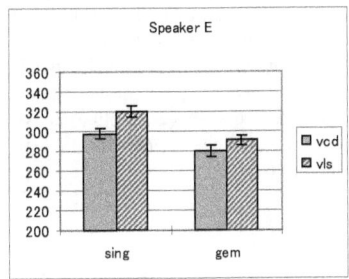
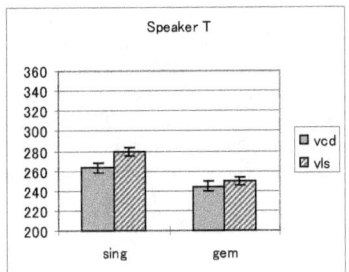
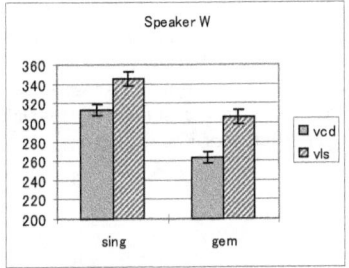

Figure 7. F0 at V2 onset (Hz).

An ANOVA shows that VOICING and LENGTH both have a statistically significant influence on F0 at V2 onset (F(1, 604)=175.945 p<.0001 and F(1, 604)=365.276, p<.0001). F0 is in general higher after voiceless consonants, although Speaker T does not show a difference after geminates (t(106)=1.341, p=.182). The interaction of VOICING and LENGTH is not significant (F(1, 604)=1.803, p=.180); however, if we look at each speaker separately, the interaction of these factors is clearly observed. Speaker E and T have a smaller difference after geminates (and in fact Speaker T's difference is lost after geminates). On the other hand, Speaker W has a larger difference after geminates (around 32Hz for singletons and 40Hz for geminates). This observation is statistically supported by the fact that the interaction of VOICING, LENGTH and SPEAKER is significant: F(2, 604)=5.387, p=.005).

Another observation is that F0 is lower after geminates (recall that LENGTH has a statistically significant impact on F0). Perhaps this is because the tonal contour of the recorded tokens is HL; given longer closure, the F0 fall is more drastic after geminates because there is more time to implement the HL fall (in a heavy syllable the fall starts at the first mora of the syllable (e.g. McCawley 1968:133-134)).

Anther point that merits discussion here is the fact that Speaker W has a larger F0 difference after geminates. This can be captured as a compensation effect in which the speaker attempts to enhance the voicing cue by F0 manipulation after geminates, whose closure voicing is weakened. Another related point is that, for Speakers E and W, the F0 difference is maintained after geminates, despite the fact that glottal vibration usually stops before release.[3] These two points suggest that manipulation of F0 is not automatic but intentional (Kingston and Diehl 1994). If it were automatic, we could not explain the fact that semi-devoiced voiced geminates have lower F0 in the following vowel. Also, the fact that a speaker can enhance an F0 difference after geminates suggests that it is possible to intentionally control F0. Finally, to the extent that this manipulation is to enhance a voicing contrast, this is in line with Kingston and Diehl's (1994) view that such manipulation is essentially to enhance phonological contrasts.

2.2.5. F0 at V2 steady state

As seen above, F0 at V2 onset is higher after voiceless consonants. F0 at V2 steady state was also measured at the sixth glottal pulse after the onset of V2 (about 10 to 20 ms away from the onset). In this position also, F0 is lower after geminates (F(1, 604)=683.250, p<.0001), and after voiced consonants (F(1, 604)=44.470, p<.0001), although the second generalization is only true for Speaker E and W, as is discussed more fully below. The interaction of LENGTH and VOICING is also significant (F(1, 604)=4.960, p=.026). This is because Speaker E and W have larger differences after geminates:

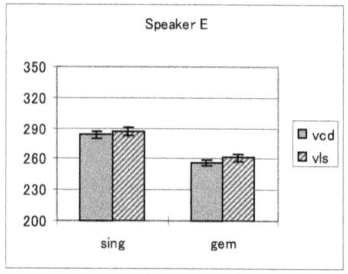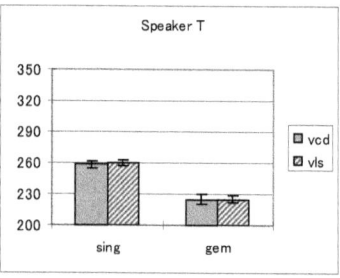

---

[3] A similar fact is reported in English [+voi] consonants where F0 depression is observed next to [+voi] consonants regardless of the presence of the actual voicing during the closure (Kingston and Diehl 1994 citing an unpublished work by Caisse (1982)).

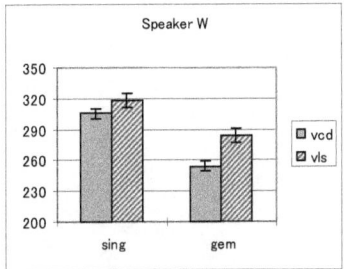

Figure 8. F0 at V2 steady state.

Looking at each speaker's behavior, Speaker T does not show any difference in terms of VOICING (t(194)=1.371, p=.172). What is more interesting is Speaker E, for whom the difference after singletons is not statistically significant, t(102)=1.166, p=.267, but the difference after geminates is, t(106)=2.529, p=.013. This pattern observed in Speaker E - that an F0 contrast emerges only after geminates - can again be captured as a compensation effect. Voicing is weakened in geminates, so that the speaker attempts to signal a voicing contrast in a way that is specific to geminates. Similarly, Speaker W has a larger difference after geminates, which can also be captured as a compensation effect. Reflecting such inter-speaker variability, the interaction of all the variables is highly significant (F(2, 604)=4.889, p=.008).

2.2.6. F0 at V1 offset

A voicing contrast is also cued by the F0 of the preceding vowel (V1), which is higher before voiceless consonants. Figure 9 illustrates the general pattern of the three speakers:

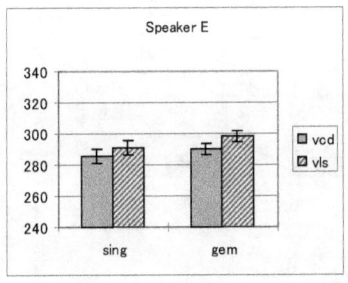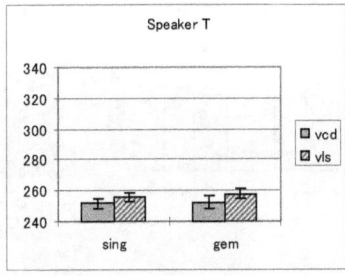

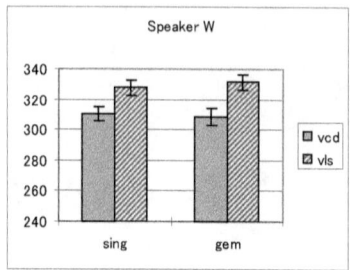

Figure 9. F0 at V1 offset.

An ANOVA shows that the influence of VOICING on F0 at V1 offset is significant: $F(1, 601) = 71.288$, $p<.0001$. A smaller main effect was observed for LENGTH: $F(1,604)=4.191$, $p=.041$. As seen in Figure 9, F0 is higher after geminates. Finally, Speaker W has a larger F0 difference than Speaker E and T, and thus the interaction of VOICING and SPEAKER is significant ($F(2, 601)=14.764$, $p<.0001$).

2.2.7. F0 at V1 steady state

The F0 values during the steady state of V1 were also measured. The measurement point was set at the sixth glottal pulse away from the offset of the vowel. There are some cases before voiceless consonants in which V1 is so short that the sixth pulse is located very close to the transitional state from the first consonant [k]. In such cases, the midpoint of the vowel was calculated, and F0 was measured at that point.

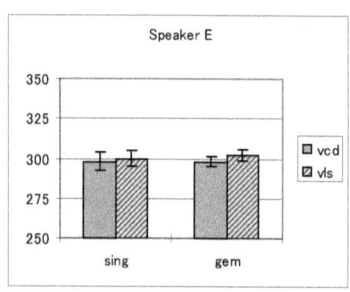 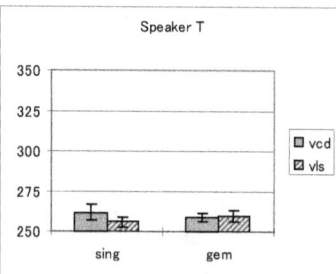

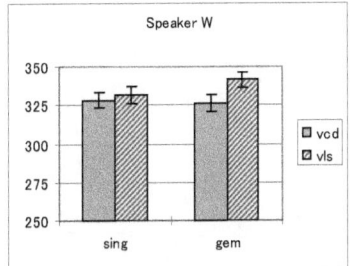

Figure 10. F0 at V1 steady state.

Overall, the difference in F0 after voiceless and voiced consonants is statistically reliable, $F(1, 599)=6.339$, $p=.012$, though not all speakers show this pattern. Speaker E has no difference before singletons ($t(100)=.375$, $p=.709$), but shows a difference before geminates (marginally significant, $t(105)=1.912$, $p=.059$). Speaker W exhibits a larger difference before geminates, while Speaker T does not show any difference either before singletons or geminates. LENGTH has no effect on F0 at the steady state of V1, $F(1, 599)=1.972$, $p=.161$. This is the tendency observed throughout the speakers; hence there is no interaction between SPEAKER and LENGTH ($F(2, 599)<1$). The interaction of VOICE and LENGTH is significant, $F(2, 599)=7.586$, $p=.006$, reflecting the fact that Speaker E and Speaker W make larger differences after geminates.

2.2.8. F1 at V2 onset

As is the case with F0, F1 is cross-linguistically known to be higher next to voiceless consonants (e.g. Kingston and Diehl 1994). To check whether such a tendency is observed in Japanese, the F1 frequency at both V2 onset and V2 steady state was measured, calculated by Praat's LPC analysis, with the number of LPC coefficients left at the default value of 10. The onset measurement point was set at the first periodic wave after the burst, and the steady state measurement point was at the sixth glottal pulse after burst.

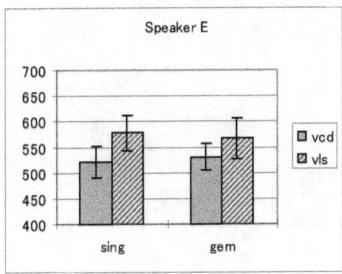 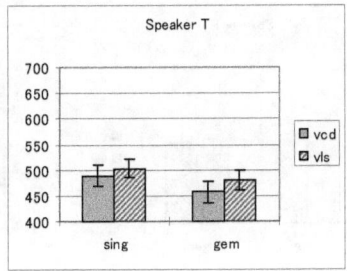

Shigeto Kawahara

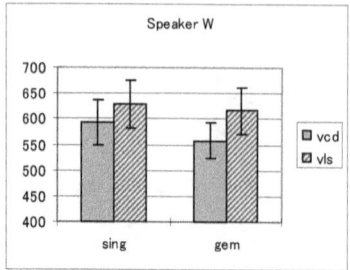

Figure 11. F1 at V2 onset. Vowel qualities are averaged over.

An ANOVA shows that VOICING affects F1 at V2 onset, F(1, 600)=14.564, p<.0001. As expected from results reported for other languages, for all the speakers, F1 is higher after voiceless consonants. LENGTH has a marginally significant effect, F(1, 600)=3.177, p=.075): F1 is lower after geminates. The size of F1 differences after voiced and voiceless consonants is similar in post-singleton and post-geminate positions; hence, no interaction of LENGTH and VOICING (F(1, 600) <1).

Here again, as was the case in F0 at V2 onset, a phonological distinction between voiceless and voiced consonants has a significant effect on F1 value, despite the fact that glottal vibration itself stops before release. This suggests that the F1 difference appearing next to voiceless/voiced consonants is not an automatic effect due to glottal vibration, but instead speakers can intentionally manipulate its values.

2.2.9. F1 at V2 steady state

F1 values during V2 steady state exhibit very consistent patterns across the speakers. Overall, both VOICING and LENGTH have a significant effect (F(1, 601)=4.378, p=.037 and F(1, 601)=12.462, p<.0001). More interestingly, no F1 differences are observed before singletons (t(287)=.118, p=.862) but a difference emerges after geminates (t(321)=2.213, p=.028). These generalizations are illustrated in Figure 12:

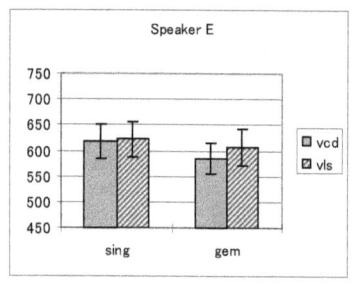
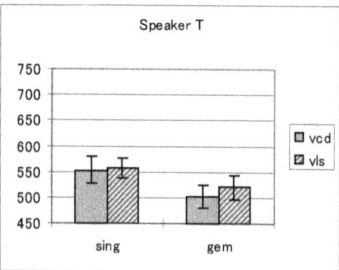

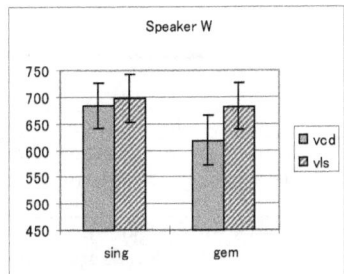

Figure 12. F1 at V2 steady state.

2.2.10. F1 at V1

The tendency for F1 to be higher next to voiceless consonants is not observed at V1, either at the offset (F(1, 600)<1) or at the steady state (F(1, 600)<1). LENGTH has an effect only at the steady state F(1, 600)=7.192, p=.008), but not at the offset (F(1, 600)<1. I do not have a good explanation on why an effect can emerge only at the steady state.

2.3.   Discussion

The purpose of the acoustic experiment described above was to see what kinds of the acoustic correlates are associated with phonemic voicing in Japanese, and how differently these acoustic correlates are realized in singletons and geminates. The experiment revealed that Japanese utilizes many of the correlates that are known to signal voicing cross-linguistically. It also showed that some phonetic differences that signal a phonemic voicing difference are attenuated in geminates (most notably closure voicing and closure duration).[4] On the other hand, the speakers attempt to compensate for the weakened cues by showing some phonetic differences between voiceless and voiced consonants only in the environment of geminates, or by making general phonetic correlates of voicing more prominent surrounding geminates. The overall results are summarized in Table 2:

---

[4] Another factor that might weaken a voicing distinction in geminates is the lack of spirantization. Voiced singletons, especially [g], spirantize whereas voiceless singletons do not, as seen in the spectrograms below:

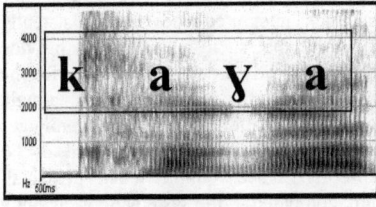 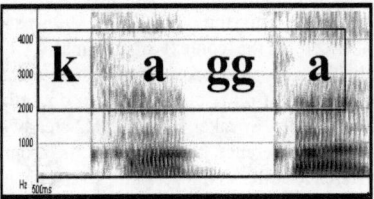

As a result of spirantization, singleton pairs like /g~/k/ are phonetically distinguished in terms of continuancy as well. However, voiced geminates do not spirantize, and therefore for geminate pairs, a continuancy difference does not signal a phonemic voicing contrast.

Phonetic cues	Change in geminates
Closure voicing	Weakened in geminates.
V1 duration	A difference is larger before geminates for Speaker T.
Closure duration	Geminate pairs are more similar to each other than singleton pairs; the vcd/vls ratio is closer to 1 in geminates.
F0 at V2 onset	A larger difference after geminates for Speaker W A smaller difference after geminates for Speakers E and T.
F0 at V2 steady state	A difference appears only after geminates for Speaker E. A larger difference after geminates for Speaker W.
F0 at V1 offset	None.
F0 at V1 steady state	A difference appears only after geminates for Speaker E. A larger difference before geminates for Speaker W
F1 at V2 onset	Speakers T and W have a larger difference before geminates.
F1 at V2 steady state	A difference emerges after geminates.

Table 2. Summary of the acoustic cues of Japanese voicing, and how they are affected by a singleton/geminate difference.

One generalization that holds for all the speakers is that closure voicing and closure duration are attenuated in geminates, but F0 and F1 differences are enhanced in one way or another surrounding geminates (modulo F0 at V2 onset for Speaker E and T).

Despite the speakers' attempt for compensation, however, it seems reasonable to speculate that overall, voicing cues are weakened in geminates. First, closure voicing, which arguably constitutes an important cue for voicing perception (Lisker 1978, 1986, Raphael 1981), is weakened in geminates. Second, the compensation effects observed above are subject to inter-speaker variation. In fact, none of the strategies is taken by all three speakers, except for the F1 difference enhancement at V2 steady state. For example, although an F0 difference at V2 onset is enhanced after geminates for Speaker W, the opposite pattern holds for Speaker E and T. Thus, unless such cues are integrated in some way (e.g. Kingston and Diehl 1995) so that such integrated cues are consistently enhanced in the context of geminates, it is doubtful that such enhancements provide reliable perceptual cues. Even if such enhancements indeed partially compensate for the weakening of other cues, it is also doubtful that the amount of compensation is enough. For example, Speaker E's F0 at V2 steady state exhibits a difference only after geminates, but the difference that emerges is around 6 Hz. Speaker W's enhancement of F0 differences after geminates at V2 onset is only 8-10 Hz. It seems unlikely that such small differences have a large perceptual effect. In sum, compared to the systematic weakening of closure voicing in geminates, the attempts for compensation are subject to inter-speaker variability, and the effects seem very small. Thus from the acoustic point of view, it seems likely that voicing cues are overall weakened in geminates. This conclusion is supported by the result of the perceptual experiment reported below in §3.

3.   Experiment II: Perceptual experiment

In order to more directly test the hypothesis that voicing is harder to hear in geminates than in singletons, a perceptual experiment was conducted. The primary aim of this experiment was to see how well Japanese speakers perceive voicing in singletons and geminates in natural environments. In order to most accurately replicate the situation in which Japanese speakers hear voicing in geminates and singletons, the natural tokens recorded in the first experiment were used. However, if I had used natural tokens and nothing else, Japanese speakers might have performed at ceiling. To overcome this problem, the stimuli were covered by cocktail party noise so as to confuse the listeners. Following the observation from the first experiment that the acoustic cues for voicing in geminates are overall attenuated, the prediction is that voicing in geminates is perceived relatively poorly compared to voicing in singletons.

3.1.   Methods

3.1.1. Stimuli

From the pool of tokens obtained in Experiment I, for each speaker, one representative example of each type of stimulus was chosen. The total number of the stimuli was therefore 108 (3 speakers × 3 vowels × 3 places of articulation × 2 consonantal lengths × 2 voicing types). Tokens that contained phonetic irregularity (such as transient sounds or devoiced V1) or spirantization were not used; for the case of singleton [g]s, which very frequently undergo spirantization (see footnote 4), tokens with least spirantization were chosen. Among the tokens of voiced geminates at each place of articulation with no phonetic distortions, those used were the ones with closure voicing duration closest to that place of articulation's average. This was in order to use representative tokens of natural voiced geminates. See the Appendix for acoustic values of the tokens used.

Cocktail party noise was used to cover the tokens. This particular kind of noise was used because to cover voicing, it was necessary to use speech-like noise that has energy in low spectra range; voicing would not be covered well by white masking noise (Miller and Nicely 1955). To obtain cocktail party noise, a party was recorded at the linguistics department of the University of Massachusetts using a SONY TCD-D8 portable DAT recorder. The recorded sound was divided into 3-second noise stretches. Six files of such stretches were randomly chosen and superimposed on top of one another. Twelve such noise files were created. To equalize the amplitudes of all the stimuli, the peak amplitudes were adjusted to 0.50 Pascal by Praat; the peak amplitudes of the noise files were modified to 0.45 Pascal. Since $dB=10\times\log_{10}(Pascal^2/.00002^2)$, the peak amplitudes of the stimuli and the noise are 87.95dB and 87.04dB, respectively. Thus the signal-to-noise ratio (S/N ratio) is 87.95dB-87.04=0.91dB (since dB is a logarithmic function, the ratio is calculated as the numerator minus the denominator). Then, one noise file was randomly chosen and was superimposed on each stimulus. After the stimuli and the noise were combined, the edges of the combined files where only noise was present were trimmed off. After this process, all stimuli were approximately 1.5 second long, including the frame sentence.

3.1.2. Subjects

In the main experiment, 15 female and 2 male native speakers of Japanese were recruited from the University of Massachusetts community. They were all in their twenties or early thirties. The speakers that participated in the first experiment were excluded since they might have some advantage hearing their own voice. All the subjects had normal hearing and were free of any speech disorders. Some were recruited from an introductory linguistic class and therefore had a basic knowledge of linguistics, but none had had extensive phonetic training. The range of dialects that the speakers spoke was diverse, including Chiba Japanese, Tokyo Japanese, Shizuoka Japanese, Ibaragi Japanese and Osaka Japanese. No report has been made of a difference in the behavior of voiced geminates among these dialects, so this dialectal variation was not expected to impact the results. Two listeners were complete bilingual speakers of Japanese and English, but their results were very similar to the results of the other subjects; hence they are included in the results reported below. All the subjects were paid or given extra credit for linguistics classes. An Informed consent form was obtained from each subject in accordance with the University of Massachusetts human research subjects guidelines.

3.1.3. Task

The experiment was conducted in a sound-attenuated booth at University of Massachusetts, Amherst. Superlab pro software (by Cedrus) was used for audio and visual presentation of each stimulus. This automatically randomizes the order of presentation. The subjects listened to stimuli over headphones (DT 250 by Beyerdynamic). They heard one stimulus at a time; as soon as a listener heard a stimulus, two choices showed up on a computer screen. The choices were minimally different in terms of voicing e.g. for [kappa], the two choices were 'kappa' and 'kabba.' The task was to make a judgment about the voicing quality based on what they heard. *Katakana* orthography was used for the visual stimuli so that people would perceive the stimuli as foreign words, in which voiced geminates are allowed. In order to make sure that speakers respond to all stimuli, there were no time limits. The listeners were not given feedback about the correctness of their response.

Before the testing sessions, they had a practice session where they did the same task for each kind of 36 tokens pronounced by one speaker. In the practice session, however, stimuli were not covered by noise, and they were given feedback about the correctness of their answers. They were also instructed to adjust the volume to a comfortable listening level during the practice session.

One testing session consisted of three blocks; each block contained all the types of stimuli pronounced by one speaker. One block thus contained 36 tokens (3 vowels × 3 places of articulation × 2 consonantal lengths × 2 voicing types)), and therefore one session contained 108 stimuli as a total. One session lasted only a few minutes. The entire experiment consisted of eight such sessions. Thus the subject heard each stimulus 24 times (3 speakers × 8 repetitions). The subjects were encouraged to take short breaks once or twice during the whole experiment. Including the instructions at the beginning and the post-experiment debriefing explanation, the entire experiment lasted about one hour.

3.2.   Results

The results of this experiment clearly show voiced geminates are misperceived as voiceless much more frequently than voiced singletons. This supports the general hypothesis of this paper that voicing is indeed harder to detect in geminates than in singletons. First, the listeners' accuracy (i.e. the proportion of correct answers across all eight trials) for each item was calculated. Averaging over the results of 17 listeners, Figure 13 summarizes the general results in terms of voicing and geminacy. This shows that voiced geminates suffer from misperception:

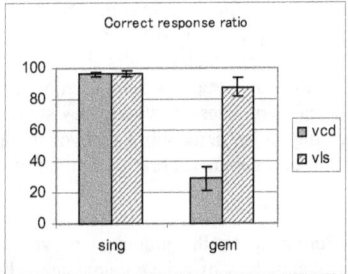

Figure 13. The average of correct response percentage out of eight trials averaged over 17 listeners.

As illustrated in the first two bars, when the target is a singleton consonant, both voiceless and voiced consonants are judged correctly more than 95 percent of the time (vls=96.4%; vcd=96.0%). Voiceless geminates are judged nearly as well (87.6%). On the other hand, voiced geminates are often misjudged: the accuracy goes down to 28.7%. This shows that voicing is indeed hard to detect in geminates, while voiced singleton consonants do not suffer from such a problem.

To verify these observations statistically, a repeated-measures ANOVA was run with VOICING (2-levels), LENGTH (2-levels), and PLACE (3-levels) as independent variables. To simplify the analysis, the two other factors (SPEAKER and VOWEL QUALITY) were averaged over. A vowel quality difference is not quite significant ($F(2,32)=3.028$, $p=.062$). Although the speaker variable exhibits a statistical difference ($F(2,32)=4.754$, $p=.016$), the mean values are not so different (Speaker E=78.5%, Speaker T=76.2, Speaker W=76.7%).

The results of the ANOVA are as follows. First, there is a large, statistically significant difference in speakers' performance between singletons and geminates consonants: $F(1, 16)=980.955$, $p <.0001$. VOICING has a main effect as well, $F(1, 16)=35.941$, $p<.0001$. These are likely to be due to the fact that voiced geminates are frequently misjudged as voiceless. This conclusion is supported by the fact that the interaction of VOICING and GEMINACY is also highly significant: $F(1,16)=45.437$, $p<.0001$. Its significance shows that voiced and voiceless consonants are judged differently in singleton and geminate context: only voiced geminates were poorly identified. No main effect is observed for PLACE ($F(2, 32)<1$, $p=.916$). Overall, the claim that voiced geminates suffer from misperception is supported.

Next, to see numerically how sensitive the Japanese speakers were to voicing in singletons and geminates, sensitivity (d') (MacMillan and Creelman 1991) was computed for each subject; d' is a measurement of sensitivity, calculated as z-transformed score of hit rates subtracted by z-transformed score of false alarm rates, where 'hit' is the probability of the listeners' correctly identifying voiced consonants as voiced, and 'false alarm' is the probability of the listeners' falsely identifying voiceless consonants as voiced.[5] A d' of zero indicates that hit and false alarm rates are the same, and that subjects have no sensitivity to voicing. The results are that the average of d' for singletons across all of the listeners is 3.794, which is significantly different from zero, $t(16)=34.15$, $p<.0001$; the average d' for geminates is 0.705, which again is significantly different from zero, $t(16)=11.47$, $p<.0001$. This shows that the Japanese listeners are sensitive to a voicing distinction both in singletons and geminates. However, they show a much higher sensitivity for singletons, again indicating higher perceptibility of voicing in singletons; a paired t-test comparing d' for singletons and geminates reveals a significant difference: $t(16)=27.27$, $p<.0001$.

Next, Figure 14 shows the listeners' performance on the judgment of voiced consonants at each place of articulation. As seen, the tendency of voiced geminates to be poorly judged holds across the three places (see below for more on differences due to PLACE):

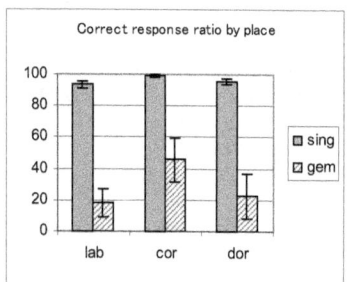

Figure 14. Correct identification rate of voiced consonants at each place of articulation.

---

[5] Since z-scores are not defined for 0 and 1, I followed MacMillan and Creelman (1991) to add or subtract the equivalent of half of one response (i.e. 1/2×n) from each perfect score. For example, if a listener identified voiceless geminates as voiceless 100 percent of the time, the proportion is 1-(1/2*216)=0.998 where 216 is the number of voiceless geminate tokens they heard.

Finally, Figure 15 illustrates the correct identification proportion for each segment type, classified according to the place of articulation. This shows that the performance of Japanese speakers to identify voicing in singletons is consistently high across all places of articulations, whereas voicing in geminates is very frequently misperceived:

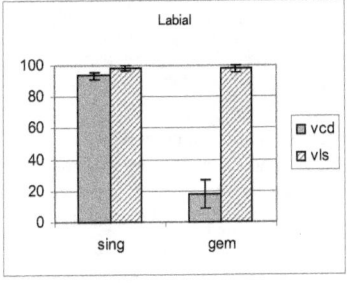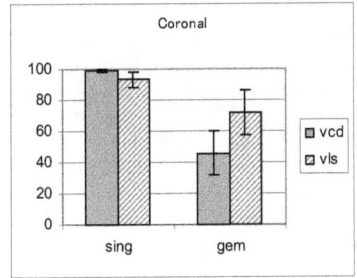

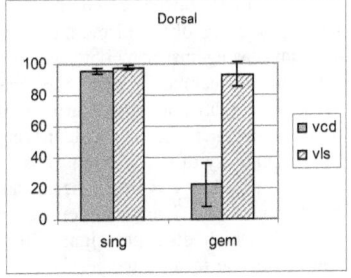

Figure 15. Correct identification proportion of each segment type at each place of articulation.

Interesting differences are observed between coronals on the one hand and labials and dorsals on the other. In labials and dorsals, voiceless geminates (as well as singletons) are judged correctly almost 100 percent of the time ([pp]=98.0%, [kk]=93.0%) while for coronals, some [tt] tokens are misheard as voiced (72%). In addition, the correct response proportion for [dd] is much higher than that for [bb] and [gg], and further [gg]'s correct response ratio is slightly higher than that of [bb] ([bb]=17.9%, [dd]=46.6% and [gg]=22.3%). To see if these differences among the three places were statistically significant, within-subject post-hoc contrast analyses were run on the proportions of correct response ratio, which revealed that there is indeed a statistically significant difference between [dd] and [gg] (t(16)=29.74, p<.0001), and between [gg] and [bb] (t(16)=4.53, p=.0002).

These perceptibility differences among the places of articulation are at least partially reflected in the likelihood of phonological devoicing. There is only one word that contains [bb] and satisfies a phonological environment in which categorical devoicing takes place (*gebberusu* 'Göbbels'), and thus it is difficult to make any conclusive generalizations about [bb]. However, analyzing Nishimura's (2003) web-based data to compare devoicing likelihood of [dd] and [gg] in the environment like (3) reveals that [gg] is more frequently transcribed as voiceless (24%; 51131 out of 216440 tokens) than [gg] (15%; 35539 out of 231752 tokens). This further supports the view that the likelihood of devoicing is closely tied to perceptibility: the harder perceptibility of voicing in [gg] results in more frequent categorical devoicing.

3.3.   Discussion

3.3.1.   Bias against perceiving voicing in geminates

The perceptual test supports the hypothesis that voicing is harder to perceive in geminates. This suggests that the weakened cues – closure voicing and closure duration – might not be compensated for by speakers' manipulation of F0 and F1 surrounding geminates (see §2.4). We can further conclude from this that closure voicing and/or closure duration are important cues to voicing in Japanese.

One interesting aspect of the results is that, listeners rarely misperceived voiceless geminates as voiced. On the other hand, voiced geminates are often misperceived as voiceless. This indicates that there might be a perceptual bias against perceiving voicing in geminates. To verify this, following MacMillan and Creelman (1991), the bias function c was calculated, which is the sum of z-scores of hit and false alarm rates multiplied by -0.5. Recall that 'hit' is the probability of the listeners' correctly identifying voiced consonants as voiced, and 'false alarm' is the probability of the listeners' falsely identifying voiceless consonants as voiced; therefore, positive c values mean that listeners had more false alarm responses than hit responses i.e. listeners preferred a voiceless response. The mean c for singletons is 0.08, which does not significantly deviate from zero ($t(16)=1.007$, $p=.33$). On the other hand, the mean c for geminates is 1.079, which is significantly different from zero ($t(16)=5.569$, $p<.0001$). This again shows that there is a perceptual bias against responding "voiced" when geminate stimuli are indeed voiced, but this is not the case for singleton stimuli. These results indicate that when Japanese speakers are confused about the voicing quality in a geminate in a noisy environment, they perceive it as voiceless by default.

To explain this observation, in addition to weakening of acoustic cues, it might be the case that lexical frequency and/or phonological constraints antagonistic against voiced geminates are also at work here. Since voicing is phonemic on geminates only in loanwords, voiced geminates are overall much less frequent than are voiced singletons. This is confirmed by a survey using Amano and Kondo's (2000) database, based on Asahi Newspaper issues from 1985 to 1998.

	Vcd	Vls	Vcd/Vls ratio
Sing	84732417(122616)	255086803 (276164)	33.2%
Gem	24587 (505)	4274451 (11792)	0.57%

Table 3. The token frequency of voiced singletons, voiced geminates, voiceless singletons and voiced geminates. The numbers in parentheses represent type frequency.

Voiced singletons are 33.2% as frequent as voiceless singletons, but voiced geminates are only 0.57% as frequent as voiceless geminates. As a consequence, voicing in geminates, which is less frequently heard, might have a disadvantage in being perceived, as frequency can cause such a perceptual bias (e.g. Hay et al. in press for a recent overview). Further, grammatical constraints antagonistic to voiced geminates might be also at work: phonologically illegal sounds or sound sequences cause a perceptual bias (e.g. Moreton 2002). From the result of the experiment *per se*, it is not clear which factor is responsible for the poor performance in judging voiced geminates (though see below for an argument in favor of a grammatical account); however, what is important is the fact that voicing in geminates is less well perceived compared to voicing in singletons, and this is perhaps related to its tendency to be easily neutralized, as argued by Kawahara (2005).

3.3.2. Place differences

The difference between coronals on the one hand and labials and dorsals on the other is noteworthy. For labials and dorsals, the Japanese speakers exhibit a bias against hearing voiced geminates; they rarely hear voiceless consonants as voiced. On the other hand, for coronals, the Japanese speakers seem to be more confused when faced with coronal geminates: as a result, the voicing judgment of coronal geminates is much closer to what is expected if people are responding by chance (in which case both [tt] and [dd] would be judged correctly 50% of the time). The calculation of the bias function c supports this observation. On average, $c(\text{cor})=0.459$, $c(\text{lab})=1.575$, and $c(\text{dors})=1.62$; labials and dorsals show more bias against voiced geminates. A within-subject contrast analysis comparing coronals and the average of labials and dorsals shows a significant difference, $t(16)=6.11$, $p<.0001$.

It can be speculated that [dd] is more acceptable for Japanese speakers so when they listen to a coronal geminate in a noisy environment, they are confused about the voicing quality, and respond more or less at chance, rather than merely rejecting the possibility of a [+voice] perception.[6] On the other hand, when they are confused with the presence of voicing in labial and dorsal geminates, listeners tend to reject the possibility of [+voice] perception.

This finding about a difference between labials and dorsals versus coronals is partially replicated by Gelbart's (2005) finding that Japanese listeners are very reluctant to hear a geminate [bb], compared to [dd] (he does not test [gg]). He reports that given a continuum of a full voiced obstruent, along which closure duration is varied, a labial is not heard as a geminate unless it has very long closure. In fact, even with the longest closure (about 200ms), a labial voiced consonant is not judged as geminate 100% of the time (a bit above 80%). In addition, the reaction time for length judgment for labial voiced consonants is longer than that for coronal consonants. This is in line with my result in that there is an extra bias against [bb] in Japanese people's perception.

That [bb] and [gg] are particularly disfavored compared to [dd] is also reflected in the lexical frequency of [dd] compared to [bb] and [gg]. According to Amano and Kondo

---

[6] After the experiment, some subjects did report that coronal geminates are harder to distinguish their voicing quality compared to others.

(2000), [dd] is much more frequent than [bb] or [gg] (based on tokens; the numbers in parentheses represent the number of types), as shown in Table 4:

Following vowel	bb	dd	gg
a	91 (4)	493(14)	142(2)
i	0	13(3)	41(6)
u	398 (6)	0	996(147)
e	0	15(4)	0
o	1 (1)	22375(316)	22(2)
SUM	490 (11)	22896 (337)	1201 (157)

Table 4. Frequency of geminate [bb], [dd], and [gg]. The numbers in parentheses stand for type frequency. The data is from Amano and Kondo (2000).

The small number of [bb] tokens reflects the fact that a labial often fails to geminate in the environment where gemination is otherwise expected (Katayama 1998; Shirai 1999): compare *knob*, borrowed as [nobu], with *dog*, borrowed as [doggu], and *God*, borrowed as [goddo].[7] From this, it seems reasonable to posit a grammatical constraint against [bb] in Japanese, which blocks gemination of [b] and hinders the perception of [bb]. Note that this blockage of gemination of [b] (and [g]) cannot be explained in terms of lexical frequency: prior to borrowing, no voiced geminates were present in the Japanese lexicon: all of [bb], [dd], and [gg] had zero frequency.[8] Therefore, some kind of grammatical constraint must have been at work. In sum, the fact that Japanese speakers do not hear [bb] and [gg] as often as they hear [dd] might suggest that there might be an additional grammatical constraint against [bb] and [gg] – [bb] and [gg] are more marked, and hence people are more reluctant to hear them. A constraint against [dd] does exist, but the requirement is not very strong; this means that people get confused between [tt] and [dd], rather than rejecting [dd].

3.3.3. Japanese-specific markedness hierarchy?

The results discussed above are inconsistent with a purely aerodynamic view of markedness, which predicts that [bb] is least marked voiced geminate (Hayes and Steraide 2004; Ohala 1983). The size of the oral cavity is biggest for labials, and cheek muscles are susceptible to passive expansion, so maintaining voicing should be easiest during [bb]. Cross-linguistically, it is indeed observed that [bb] is more frequent than [dd] (see the works cited above). There must thus be a Japanese-specific constraint against [bb]. The reason for this language-specific behavior of Japanese is yet to be explained.

The higher markedness of [gg] over [dd] might be attributable to the general marked scale *[gg] » *[dd], which derives from aerodynamics for the same reasons

---

[7] [g] is less likely to undergo gemination than [d], although it does not resist gemination as much as [b].

[8] One exception is the resulting forms of emphatic gemination, e.g. *suggoi* from *sugoi* 'formidable'. See Kawahara (2001) for a discussion of the non-structure preserving nature of this process.

described above. The size of oral cavity is smaller for [gg] than for [dd], and its capacity to actively and passively expand is much smaller (see the references above). Therefore, it is easier to maintain voicing in coronals than dorsals, hence the extra bias against [gg].

4. Conclusion

This paper has investigated acoustic and perceptual aspects of voicing in Japanese geminates. As Japanese only recently phonemicized voicing in geminates, little work has been done on the interaction of voicing and geminacy. I have identified various phonetic cues that signal voicing in Japanese: (i) closure voicing, (ii) closure duration, (iii) V1 duration (iv) F0 and F1 of surrounding vowels. Some of these cues are weakened in the context of geminates. Most notably, closure voicing is weakened in geminates because maintaining glottal vibration during a long obstruent closure is aerodynamically hard. On the other hand, we have also seen that some other cues are enhanced in the context of geminates, perhaps to compensate for the weakening of other cues. Compared to the systematic weakening of closure voicing, however, such attempts for compensation are subject to inter-speaker variability, and the effects seem small. So from the acoustic point of view, it seems likely that voicing cues are overall weakened in geminates.

This conclusion has been supported by the results of the perceptual experiment. The experiment revealed that voicing is much harder to discriminate in geminates in general, supporting the hypothesis advanced in Kawahara (2005). The experiment shows that weakened cues are important for voicing perception in Japanese, and the attempt for compensation is not sufficient. What has not been investigated in this research, however, is which acoustic correlates contribute to voicing perception to what extent. This is a topic for future research.

Furthermore, I pointed out that there is a possibility that, in addition to weakening of acoustic cues, Japanese grammar might have constraints that yield a perceptual bias against labial and dorsal voiced geminates, which also manifest themselves through the patterns of gemination. This explains the particularly poor performance in identifying voicing in [bb] and [gg], as well as the fact that Japanese listeners almost never misidentified voiceless geminates as voiced. This provides another case in which grammatical constraints may affect people's perception (e.g. Moreton 2002). Lexical frequency might also be at work, though it fails to explain the blockage of gemination in loanwords. One remaining question is why [dd] is more favored than [bb], despite that aerodynamically [bb] is predicted to be more unmarked than [dd].

One general conclusion that can be drawn from the experiments reported in this paper is that phonology is at least partially driven by phonetics. Phonologically, voicing is more easily lost in geminates than in singletons in Japanese. In light of the result of this experiment – that voicing is harder to hear in geminates – we can regard this as another case in which contrasts signaled by phonetically weak cues are phonologically more prone to neutralization, just as preconsonantal place cues are much more easily lost than prevocalic place cues. This finding adds to the growing body of literature that shows phonological neutralization is closely tied to phonetic perceptibility (Hura et al. 1992; Jun 2004; Kohler 1990; Steriade 2001). This study thus provides an additional endorsement of the claim that phonology is, at least in part, affected by phonetic factors.

Shigeto Kawahara

Appendix. Acoustic values of the tokens used in Experiment II

				11			
	voicing duration (ms)	closure duration (ms)	V1 duration (ms)	F0 at V1 (Hz)	F1 at V1 (Hz)	F0 at V2 (Hz)	F1 at V2 (Hz)
kapa	15	54	29	286	825	311	768
kepe	15	72	35	309	533	308	546
kopo	19	63	25	290	594	306	574
kaba	57	57	39	297	714	284	723
kebe	45	45	38	301	495	298	530
kobo	45	45	31	292	420	296	502
kappa	16	132	50	295	809	259	796
keppe	12	127	52	289	542	273	539
koppo	17	133	52	304	590	278	519
kabba	31	93	50	271	811	266	742
kebbe	36	113	72	308	571	267	512
kobbo	44	123	55	285	535	255	513
kata	17	62	30	300	660	320	621
kete	17	63	42	311	516	317	515
koto	24	63	38	295	443	309	489
kada	61	61	43	273	511	288	575
kede	46	46	51	283	354	284	454
kodo	43	43	38	283	421	288	462
katta	2	134	62	299	595	261	555
kette	29	159	62	290	431	312	480
kotto	24	146	62	288	471	267	475
kadda	41	105	74	295	601	284	588
kedde	37	121	76	285	464	266	448
koddo	45	149	79	169	428	274	457
kaka	8	39	33	290	613	321	781
keke	8	42	43	298	338	346	418
koko	2	55	41	295	458	331	538
kaga	36	36	42	286	483	279	630
kege	43	43	73	285	296	287	321
kogo	51	51	58	308	421	288	481
kakka	9	129	63	308	699	263	790
kekke	5	115	69	334	374	284	399
kokko	15	120	51	327	466	273	531
kagga	34	89	61	304	609	302	592
kegge	52	130	83	287	333	285	392
koggo	40	92	52	304	471	265	527

			Speaker T				
	voicing duration (ms)	closure duration (ms)	V1 duration (ms)	F0 at V1 (Hz)	F1 at V1 (Hz)	F0 at V2 (Hz)	F1 at V2 (Hz)
kapa	6	74	33	270	439	283	569
kepe	17	81	52	248	449	282	498
kopo	21	66	27	250	480	266	482

kaba	43	43	31	257	561	207	547
kebe	46	46	21	244	386	259	429
kobo	46	46	34	255	504	258	502
kappa	17	115	53	259	532	249	542
keppe	16	136	59	251	494	243	505
koppo	14	131	56	264	484	254	478
kabba	30	139	83	354	505	219	574
kebbe	38	141	84	263	410	247	471
kobbo	42	120	51	249	488	237	454
kata	0	62	30	259	595	276	582
kete	11	88	50	255	459	254	442
koto	8	55	41	258	512	284	511
kada	26	26	40	258	600	281	538
kede	38	38	63	252	450	252	450
kodo	32	32	50	278	502	273	460
katta	21	107	58	259	600	284	564
kette	14	115	68	268	536	252	472
kotto	0	144	42	257	494	277	469
kadda	40	103	74	266	613	260	528
kedde	35	93	98	262	515	249	457
koddo	24	102	90	270	523	238	465
kaka	10	70	42	255	543	278	565
keke	9	52	49	251	485	297	472
koko	0	51	58	249	431	276	512
kaga	41	41	60	253	503	268	526
kege	39	39	85	254	339	257	370
kogo	52	52	55	244	360	249	319
kakka	7	126	54	260	541	252	548
kekke	0	130	59	241	462	240	456
kokko	9	140	52	260	467	246	457
kagga	41	139	98	242	418	212	504
kegge	46	122	73	243	424	225	427
koggo	44	122	88	261	354	239	403

Speaker W						
	voicing duration (ms)	closure duration (ms)	V1 duration (ms)	F0 at V1 (Hz)	F1 at V1 (Hz)	F0 at V2 (Hz)	F1 at V2 (Hz)
kapa	2	95	33	295	828	275	820
kepe	13	77	36	344	641	335	610
kopo	32	76	26	333	642	339	488
kaba	53	53	47	313	933	320	906
kebe	50	50	64	337	623	339	593
kobo	46	46	39	321	466	338	440
kappa	19	147	47	287	855	268	776
keppe	11	149	56	328	573	286	536
koppo	10	136	61	330	388	310	553
kabba	23	121	63	300	879	259	769
kebbe	37	87	104	234	661	294	409
kobbo	36	115	85	328	670	252	474

Shigeto Kawahara

kata	0	81	25	288	848	270	810
kete	13	61	52	318	501	335	523
koto	11	62	42	336	557	346	597
kada	33	33	42	305	857	299	730
kede	38	38	62	304	443	314	400
kodo	31	31	71	275	491	266	495
katta	7	142	34	314	893	275	795
kette	8	130	66	328	461	307	482
kotto	0	138	66	317	470	295	566
kadda	35	121	39	307	917	265	820
kedde	34	116	79	318	377	264	477
koddo	35	116	74	349	647	281	525
kaka	0	56	34	343	666	359	951
keke	0	56	47	335	442	343	406
koko	11	66	46	340	566	356	618
kaga	40	40	59	307	594	324	761
kege	29	29	88	299	411	286	410
kogo	44	44	58	280	403	272	501
kakka	0	120	38	320	777	288	907
kekke	11	124	61	353	402	336	419
kokko	12	123	59	322	443	306	533
kagga	37	116	76	323	681	262	665
kegge	27	77	89	335	771	317	428
koggo	43	132	84	328	409	277	511

References

Amano, Shigeaki and Tadahisa Kondo (2000) *NTT database Series: Lexical Properties of Japanese, 2nd Release.* Tokyo: Sanseido.

Boersma, Paul, and David Weenink (1992) *Praat: Doing Phonetics by Computer.*

Chen, Matthew (1970) Vowel length variation as a function of the voicing of the consonant environment. *Phonetica* 22:129-159.

Fougeron, Cecile and Patricia Keating (1997) Articulatory strengthening at edges of prosodic domains. *Journal of the Acoustical Society of America* 106: 3728-3740.

Gelbart, Ben (2005) *Perception of Foreignness.* Doctoral dissertation, University of Massachusetts, Amherst.

Hay, Jennifer, Janet Pierrehumbert and Mary Beckman (in press) Speech perception, well-formedness, and the statistics of the Lexicon. To appear in *Papers in Laboratory Phonology VI.* Cambridge: Cambridge University Press.

Hayes, Bruce and Donca Steriade (2004) Introduction: The phonetic bases of phonological markedness. In B. Hayes, R. Kirchner, and D. Steriade (eds.) *Phonetically-based Phonology.* Cambridge: Cambridge University Press. pp. 1-33.

Hura, Susan, Björn Lindblom and Randy Diehl (1992) On the role of perception in shaping phonological assimilation rules. *Language and Speech* 35: 59-72.

Itô, Junko and Armin Mester (1996) Stem and word in Sino-Japanese. In T. Otake and A. Cutler (eds.) *Phonological Structure and Language Processing: Cross-Linguistic Studies.* Berlin: Mouton de Gruyter. pp. 13-44.

Itô, Junko and Armin Mester (1999) The phonological lexicon. In N. Tsujimura. (ed.) *The*

Handbook of Japanese Linguistics. Oxford: Blackwell. pp. 62-100.

Itô, Junko and Armin Mester (2003) *Japanese Morphophonemics*. Cambridge: MIT Press.

Jaeger, Jeri (1978) Speech aerodynamics and phonological universals. *Proceedings of Chicago Linguistics Society* 311-329.

Jun, Jongho. (2004) Place assimilation. In B. Hayes, R. Kirchner, and D. Steriade (eds.) *Phonetically-based Phonology*. Cambridge: Cambridge University Press. pp. 35-57.

Katayama, Motoko (1998) *Loanword Phonology in Japanese and Optimality Theory*. Doctoral dissertation, University of California, Santa Cruz.

Kawahara, Shigeto (2001) *Similarity among Variants: Output-Variant Faithfulness*. BA Thesis, International Christian University.

Kawahara, Shigeto (2005) A faithfulness scale projected from phonetic perceptibility: The case of voicing in Japanese. Ms., University of Massachusetts, Amherst.

Kingston, John and Randy Diehl (1994) Phonetic knowledge. *Language* 70: 419-455.

Kingston, John and Randy Diehl (1995) Intermediate properties in the perception of distinctive feature values. In B. Connell and A. Arvaniti (eds.) *Phonology and Phonetic Evidence: Papers in Laboratory Phonology IV*. Cambridge: Cambridge University Press. pp. 7-27.

Kohler, Klaus. J. (1990) Segmental reduction in connected speech: Phonological facts and phonetic explanation. In W.J. Hardcastle and A. Marchals (eds.) *Speech Production and Speech Modeling*. Dordrecht: Kluwer Publishers. pp. 69-92.

Kuroda, Shige-Yuki (1965) *Generative Grammatical Studies in the Japanese Language*. Doctoral Dissertation, MIT.

Lisker, Leigh (1957) Closure duration and the intervocalic voiced-voiceless distinction in English. *Language* 33: 42-49.

Lisker, Leigh (1978) On buzzing the English /b/. *Haskins Laboratories Status Report on Speech Research SR-55/56*: 251-259

Lisker, Leigh (1981) On generalizing the rapid-rabid distinction based on silent gap duration. *Haskins Laboratories Status Report on Speech Research SR-65*: 251-259

Lisker, Leigh (1986) "Voicing" in English: A catalog of acoustic features signaling /b/ versus /p/ in trochees. *Language and Speech* 29: 3-11.

Lovins, Julie (1973) *Loanwords and the Phonological Structure*. Doctoral dissertation, University of Chicago.

Macmillian, Neil and Douglas Creelman (1991) *Detection Theory: A User's Guide*. Cambridge: Cambridge University Press.

Maddieson, Ian (1985) Phonetic cues to syllabification. In V. Fromkin (ed.) *Phonetic Linguistics*. London: Academic Press. pp. 203-221

McCawley, James (1968) *The Phonological Component of a Grammar of Japanese*. Hague: Mouton.

Miller, George and Patricia Nicely (1955) An analysis of perceptual confusions among some English consonants. *Journal of Acoustical Society of America* 27: 338-352.

Moreton, Elliot (2002) Structural constraints in the perception of English stop-sonorant clusters. *Cognition* 84: 55-71.

Nishimura, Kohei (2001) Lyman's Law in loanwords. Ms., University of Tokyo.

Ohala, John (1983) The origin of sound patterns in vocal tract constraints. In P. MacNeilage (ed.) *The Production of Speech*. New York: Springer Verlag. pp 189-216.

Poser, William (1990) Evidence for foot structure in Japanese. *Language* 66: 78-105.

Raphael, Lawrence (1972) Preceding vowel duration as a cue to perception of the voicing characteristic of word-final consonants in American English. *Journal of Acoustic Society of America* 51: 1296-1303.

Raphael, Lawrence (1981) Duration and contexts as cues to word-final cognate opposition in English. *Phonetica* 38: 126-47.

Shirai, Setsuko (1999) *Gemination in Japanese Loan Words.* MA Thesis, University of Washington.

Slis, Imas (1986) Assimilation of voice in Dutch as a function of stress, word boundaries and sex of speaker and listener. *Journal of Phonetics* 14: 311-326.

Smith, Caroline (1995) Prosodic patterns in the coordination of vowel and consonant gestures. In B. Connell and A. Arvaniti (eds.) *Phonology and Phonetic Evidence: Papers in laboratory phonology IV.* Cambridge: Cambridge University Press. pp. 205-222.

Steriade, Donca (2001) The phonology of perceptibility effect: The P-map and its consequences for constraint organization. Ms., University of California, Los Angeles.

Stevens, Kenneth and Sheila Blumstein (1981) The search for invariant acoustic correlates of phonetic features. In P.D. Eimas and J.L. Millers (eds.) *Perspectives on Study of Speech.* Hillsdale: Erlbaum. pp. 1-38.

Takagi, Naoyuki and Virginia Mann (1994) A perceptual basis for the systematic phonological correspondences between Japanese loan words and their English source words. *Journal of Phonetics* 22: 343-356.

Westbury, John (1979) *Aspects of the Temporal Control of Voicing in Consonant Clusters in English.* Doctoral Dissertation, University of Texas, Austin.

Department of Linguistics
South College
University of Massachusetts, Amherst
Amherst, MA 01003

kawahara@linguist.umass.edu

The perception of epenthetic stops in English:
The effects of cluster type and silent interval duration[*]

Takahito Shinya

University of Massachusetts, Amherst

1.   Introduction

Speech abounds with different kinds of variability. Generally, there are two types: phonological and phonetic variability. A speech sound may be phonologically altered in one way or another (assimilation or dissimilation), dropped (deletion) or inserted (epenthesis) in various circumstances. Such phonological variability is always predictable since underlying forms are recoverable from their contexts. Phonetic variability, on the other hand, involves variation that is not predictable. For example, English word-final nasals may, but not always, assimilate to the place of the following initial consonant (Gaskell and Marslen-Wilson 1996). Dutch has an optional epenthetic schwa as in [tylp]~[tyləp] (Donselaar, Kuijpers and Culter 1999). In this paper we address the latter type of variability.

This paper deals with epenthetic stops that are inserted between nasals or laterals and voiceless obstruents in English. Some examples are given in (1a), contrasting with underlying stops in similar or the same environments (1b). Epenthesized stops are indicated by superscripted symbols.

[*] I thank Lyn Frazier, John Kingston and Tom Roeper for their advice on this study. Also thanks to the editors of this volume Kathryn Flack and Shigeto Kawahara for their comments on an earlier draft. All remaining errors and shortcomings are, of course, my own.

Kathryn Flack and Shigeto Kawahara (eds.), UMOP 31, 121-149.

(1)   Epenthetic vs. underlying stops in English
        a.   Epenthetic   b.   Underlying
              [pɹɪnˢs]   prince   [pɹɪnts]   prints
              [fɔlˢs]   false   [fɔlts]   faults
              [hæmᵖstɚ]   hamster   [dʌmpstɚ]   dumpster
              [dɹɛmᵖt]   dreamt   [tɛmpt]   tempt, temped
              [jʌŋᵏstə]   youngster   [pɹæŋkstɚ]   prankster

        The phenomenon of stop epenthesis is well-known in English; epenthesized stops apparently neutralize the underlying distinction between a sonorant followed by an obstruent (e.g. /ns/) and a sonorant followed by a stop followed by an obstruent (e.g. /nts/); both of these forms have the identical surface forms (e.g. [nts]). Therefore, words such as *prince* vs. *prints* and *false* vs. *faults* may not be distinguished.

        This study focuses on perceptual aspects of epenthetic stops in English. We ask the following three questions:

(2)   a. Is it phonological constraint or cluster frequency (or both) that is at play in the perception of epenthetic stops?
         b. Under what circumstances do epenthetic stops (or more generally segments) are perceived?
         c. In identifying a stop, how are the acoustic properties of the adjacent segments used?

We see each of these questions in turn below.

        The primary question we ask is whether or not phonological constraints influence the perception of epenthetic stops. One suggestion made by Warner and Weber (2001), who experimentally examined the perception of epenthetic stops in Dutch, is that phonotactics are an influential factor. They found that stops are less likely to be perceived as underlying between nasals that are followed by obstruents when doing so violates a phonotactic constraint of the language (e.g. more /t/ percepts in /ns/ than in /np/, where /nts/ is legal but */ntp/ is not). In a related study (Warner and Weber 2002), they present statistical data on cluster frequency and suggest that frequency may also affect the perception of epenthetic stops, such that they may be perceived more in frequent clusters than in infrequent ones. Thus, the specific question that this paper addresses is: is it phonological knowledge or frequency that is at play in the perception of epenthetic stops?

        A related question in this paper is under what environment segments are perceived. In other words, we ask what acoustic properties would serve as the cues for a segment percept. Some studies, e.g. Ali, Daniloff and Hammarberg (1979) and Warner and Weber (2001), point out that in the perception of naturally produced nasal-stop/fricative clusters, listeners often hear a stop even if there is no acoustic burst between the two consonants. This suggests that a burst is not a necessary condition for perceiving a stop. Some other acoustic properties should play a role. One such property would be silent gaps between sonorants and obstruents. In the experiment reported in this paper, we examine how a silent interval of varying duration affects the perception of stops between different sonorants and fricatives.

Another question addressed in this paper is, when one hears a stop where there is only a silent interval, how would one use the acoustic properties of the adjacent segments to identify this stop? Clements (1987) claims that an epenthetic stop between a sonorant and a fricative shares its [place] feature with the preceding sonorant and its [voice] feature with the following fricative. However, bursts and formant transitions which cue epenthetic stops are often present in natural speech, which hinders us from investigating listeners' use of the acoustic properties of the adjacent segments themselves. Below we report a perception experiment in which only a silent interval of varying duration between sonorants and voiceless fricatives induces stop percepts. Thus, the acoustic property that cues the stops was only the silent interval, with no burst or formant transition contained in the stimuli. Therefore, the only information that can be used to identify the stop is the acoustic properties of the adjacent sonorant and fricative.

We report a perceptual experiment that examines the effect of cluster type and silent interval duration. We test several cluster types each of which consists of a sonorant and a fricative. A silent interval of varying duration is inserted between the sonorant and the fricative. We show that although phonological constraints and cluster frequency both do not explain the obtained data quite satisfactorily, the phonology account is superior to the frequency account for explaining one cluster type. We also show that silent interval plays an important role in the perception of epenthetic stops, which suggests that a stop burst is not a necessary condition to perceive an epenthetic stop. Furthermore, we see that the featural composition of an epenthetic stop does not always follow what Clements (1987) claims. Specifically, we see a case in which a perceived epenthetic stop shares its place feature with the following obstruent and not with the preceding sonorant.

In the rest of the paper, We first review the previous studies on epenthetic stops (§2) and then describe two different general approaches that have been proposed in the literature to account for the phonotactic effect found by Warner and Weber (2001): one based on phonological knowledge on syllable structure constraints and one based on cluster frequency (§3). The report of the experiment follows in §4. §5 concludes the paper.

2.   Previous studies on epenthetic stops

Epenthetic stops in English have a number of characteristics. Clements (1987) argues that an epenthetic stop always shares the [voice] feature with the following obstruent and the [place] feature with the previous sonorant. Epenthetic stops normally arise when the obstruent is voiceless, but, though much less commonly, they also appear when it is voiced (e.g. /nz/).(Fourakis and Port 1986).

Epenthetic stops show another interesting property. Jones (1966) and Fourakis and Port (1986) claim that stop epenthesis in English is dialect-specific. Jones (1966), based on his impressionistic observations, points out that the occurrence of an epenthetic stop between a sonorant and a fricative is not characteristic of British English. Fourakis and Port (1986) experimentally showed that epenthetic stops occur in American English but never in South African English.

While stop epenthesis has been shown to be dialect-specific, there is also an opposing idea that the phenomenon is governed by articulatory constraints and is universal to all languages. Ohala (1974) suggests that in a nasal-obstruent sequence a stop is produced unintentionally in the articulatory transition from the nasal to the following obstruent, due to mistiming of the closure of the velic port and the oral release of the nasal. Specifically, if the raising of the velum for the velic closure for the following obstruent occurs before (instead of simultaneously with) the oral release at the offset of the nasal, the oral cavity is sealed, and a burst will arise when the oral closure for the nasal is released. Ohala (1974) also suggests that the occurrence of stops in lateral-fricative clusters is explained in terms of universal constraints on articulation. Based on palatographic data, he points out that the contact areas for /l/ and /s/ are to a certain degree complementary. He says that "in moving from an [l] to an [s], contact and release of contact must be made simultaneously in these complementary areas" and that "to the extent that they are not simultaneous, complete contact all around the alveolar ridge may result and thus complete stoppage of the air, that is, a [t] will result" (Ohala 1974: p. 359).

Ali, Daniloff, and Hammarberg (1979) is the first study that systematically examined nasal-fricative clusters in English, using oral and nasal air flow and oral air pressure data. They found that what happens first in producing an epenthetic stop is cessation of voicing during the nasal, followed by cessation of the nasal airflow. Because voicing and nasal airflow cease, the oral air pressure increases and reaches its maximum toward the offset of the nasal, while the oral closure is maintained. When the oral closure for the nasal is released, a burst appears due to the abrupt emission of the highly pressured air from the oral cavity.

Ali et al. (1979) also provide preliminary perception data obtained by auditory judgments from 3 listeners. They report a finding that some /ns/ clusters elicited epenthetic stop perception even without a burst or silent gap being present. They also note that the presence of a syllable or morphological boundary between the nasal and the fricative does not elicit perception of epenthetic stops. This is consistent with the results obtained by Warner and Weber (2001), which we review below.

Fourakis and Port (1986) did a production experiment on the English /ns/, /nz/, /ls/ and /lz/ clusters in monosyllabic word-final position. They analyzed data on American English (midwestern) and South African English, and claimed that stop epenthesis is a dialect-specific phenomenon that is observed only in American English. They found that, in American English, when the fricative was voiceless there was always an interval of silence lasting more than 10 ms between the sonorant and fricative, which they judged as the occurrence of an epenthetic stop. When the fricative was voiced, epenthetic stops were still observed, but far less often than when the fricative was voiceless. No such silence was seen for South African English in either of the conditions. Based on these findings, Fourakis and Port argue against the claim that epenthetic stops are caused by universal constraints on articulation and aerodynamics.

Fourakis and Port's (1986) claim that stop epenthesis is dialect-specific appears to be a bit too strong, because they base their claim on a purely acoustic parameter: the presence of silent intervals of more than a certain duration. Though there is a systematic difference between American English and South African English in articulating sonorant-obstruent clusters, we need perceptual evidence that a 10 ms silent interval in /ns/ and /ls/ is always identified as a stop in order to make their claim valid.

Fourakis and Port (1986) also found a systematic temporal difference between clusters with epenthetic stops and clusters with underlying stops. They show that when the stop is underlying as in /nts/ or /ndz/, the oral closure duration is longer and the vowel and nasal are both shorter than when the stop is epenthetic. When the sonorant is /l/ as in *false* and *faults* the vocalic portion /ɔl/ is shorter when it is followed by the underlying /t/ than by the epenthetic [t]. This finding is important because it raises an interesting question of whether listeners use these acoustic differences as cues for distinguishing between epenthetic and underlying stops.

However, the results of the subsequent studies on closure duration of epenthetic stops are mixed. In her study of the English /ns/ cluster using the TIMIT database (see Keating, Blankenship, Byrd, Flemming, and Todaka 1992 for the description of the database), Blankenship (1992) showed that epenthesis occurs in about one-fourth of the /ns/ strings of American English. Though she found that epenthetic stops were shorter than underlying stops, the difference was not statistically significant.

Yoo and Blankenship (2003) recently followed up these results with an independent production experiment. They examined the difference in stop closure duration between the /ns/ and /nts/ clusters in American English appearing in different positions in a syllable and in different stress environments. They confirmed Blankenship's (1992) finding that no durational difference was found between underlying and epenthetic stops, with one exception. They found a statistically significant difference in the environment where the /ns/ cluster was separated by a syllable boundary and the following vowel was stressed, as in *consent* vs. *blunt saying*. The closure durations in the /ns/ and /nts/ clusters did not differ significantly in word-final position (e.g. *intense* vs. *intents*), contra Fourakis and Port's (1986) result. Also, it turned out that stress does not affect the closure duration, since Yoo and Blankenship (2003) did not find statistically significant effects of stress.

Yoo and Blankenship (2003) also reanalyzed the occurrence of the /ns/ cluster using the TIMIT database with respect to its position in a word and stress, and obtained a result that is different from Blankenship's (1992). They found that the closure duration was significantly shorter for the /ns/ cluster than for the /nts/ cluster in any position and in any stress condition.[1] The issue still remains to be settled.

Warner and Weber (2001) experimentally investigated the perceptual characteristics of the epenthetic stops that appear between nasals /m, n, ŋ/ and voiceless stops /p, t, k/ or the fricative /s/ in Dutch. They recorded two speakers who naturally

---

[1] Yoo and Blankenship (2003) do not explain why this discrepancy was found. The reason is not at all clear.

produced monosyllabic nonwords containing all of the clusters obtained by combining the nasals and the obstruents in word-final position. The speakers were explicitly instructed to produce nasals without assimilating to the places of the following obstruents. Warner and Weber first analyzed the production data acoustically. The data show considerably different patterns with respect to the presence of burst depending on the cluster type.[2] Specifically, very few bursts were observed in the clusters such as /np/ and /ŋp/, while relatively high rates of epenthetic burst occurrences were seen in the other clusters. Warner and Weber offer an articulatory speculation. They suggest that in a cluster with the nasal posterior to the intended stop like /np/ or /ŋp/, the oral closure for the stop may often be made before the release of the oral closure for the nasal, masking the latter articulatory gesture. Therefore, no burst results.

Warner and Weber (2001) presented their listeners with the recorded materials whose acoustic characteristics have just been described above in a phoneme monitoring task (see Connie and Titone, 1996 for a review of the method). The listeners were asked to monitor the kind of epenthetic stop that is expected in the particular cluster. For example, they were required to monitor /p/ if the stimulus was /zymt/. Warner and Weber found that listeners perceived epenthetic stops about 50% of the time. They also showed that listeners tend to miss an epenthetic stop more often if it would form a phonotactically illegal cluster with the flanking consonants than if it would not. In Dutch, heterorganic two-consonant clusters are illegal, except when the second one is coronal. Thus, in combining the nasals /m, n, ŋ/ and the voiceless stops /p, t, k, s/, only /mt, ms, ns, ŋt, ŋs/[3] are legal while the others /mk, np, nk, ŋp/ are illegal. In three-consonant clusters, only clusters with two homorganic consonants followed by a coronal are legal. Thus, the same legal and illegal sets of clusters seen in two-consonant clusters can be extended to three-consonant clusters: /mpt, mps, nts, ŋkt, ŋkt/ are legal whereas the others /mpk, ntp, ntk, ŋkp/ are illegal. Warner and Weber's listeners perceived more epenthetic stops when they heard the legal clusters such as /mt/ and /ŋs/ than the illegal ones such as /mk/ and /nk/, because /mpt/ and /ŋks/ are legal in Dutch but /mpk/ and /ntk/ are not. Furthermore, they found that listeners were less likely to respond to epenthetic [t] than to epenthetic [p] or [k]. Interestingly, in the environments where an epenthetic [p] or [k] appears, listeners often heard a stop even when a burst was not present, but in those where an epenthetic [t] appears, they did not hear a stop unless there was a burst in the signal. Warner and Weber first relate this effect to the widely acknowledged special status of coronals. However, based on the experimental result in Hume, Johnson, Seo, and Tserdanelis (1999) that

---

[2] Warner and Weber (2001) adopted the criterion of whether an acoustic burst appears or not, rather than whether there is a silent interval of a certain duration or not, in determining the presence of an epenthetic stop. They suggest that "it is not always possible to determine from acoustic information that a silent gap is a reflection of an epenthetic stop" and "in clusters such as /ŋp/, an epenthetic /k/ might be released while the labial closure is already being made, thus preventing the velar burst from appearing in the acoustic record" (p. 61). In this case a silent gap is produced by an articulatory mechanism that is different from the one that causes a gap in clusters like /mt/, where a gap is produced by the cessation of voicing while the oral closure of the nasal is being made.

[3] Clusters /mp, nt, ŋk/ are omitted. Though they are legal in Dutch, no epenthetic stop is expected in any of them since the two consonants are homorganic, which means that articulatory contact holds throughout the clusters.

coronal place is only slightly less salient than other places when stops are released, they conclude that the low rate of perception of epenthetic /t/ in their experiment derives from the combination of several factors which do not involve the special status of coronals. They attribute the effect to factors such as orthographic and production influences.

In their subsequent study, Warner and Weber (2002) investigated the effect of a syllable boundary on the perception of epenthetic stops in Dutch. They replicated Ali et al.'s (1979) result that listeners hear fewer epenthetic stops when the nasal and the following obstruent are separated by a syllable boundary. Furthermore, Warner and Weber (2002) conducted a statistical survey on the frequency of the nasal-stop/fricative clusters in Dutch and English, using the CELEX database (Baayen, Piepenbrock, and van Rijin 1993). They found that, both within a syllable and between syllables, clusters in which epenthetic stops potentially arise are far more frequent in Dutch than in English, and suggest that epenthetic stops are predicted to be observed more in Dutch than in English.

We have seen above that one of the major findings of Warner and Weber's (2001) study is the phonotactic effect. They note that "listeners are more likely to interpret an unintended epenthetic stop as an occurrence of the stop phoneme if doing so does not violate a syllable structure constraint in their language." (p.78). In this view, the phonotactic effect is accounted for by assuming phonological knowledge in the form of abstract rules or constraints that prohibit particular sequences of phonemes (Moreton 2002, Moreton and Amano 1999). For example, the illegality of the cluster */ŋkp/ in Dutch, as opposed to /ŋkt/, can be explained in such a way that there is a constraint that bans segments that have noncoronal places of articulation in the *syllable appendix*, the subsyllabic unit that does not belong to the rhyme but is directly adjoined to the syllable node (Booij 1995).

However, I argue that Warner and Weber's (2001) claim is invalid. In Dutch, it is always the case, at least with the clusters that Warner and Weber tested, that when a two consonant cluster is illegal (e.g. */mk/) the corresponding three consonant cluster is also illegal (e.g. */mpk/). Similarly, when a two consonant cluster is legal (e.g. /mt/), then the corresponding three consonant cluster is legal too (e.g. /mpt/). Thus no perceptual bias is expected toward one of the two clusters if both are illegal (or legal). The results obtained by Warner and Weber (2001) must be accounted for by some other principles. The phonotactic effect would be observed with respect to epenthetic stops when only one of the clusters is illegal. In that case, listeners would be biased toward the legal cluster.

Even in such cases, however, an alternative account based on cluster frequency is possible. There is large body of work in the literature suggesting that frequency affects the perception of language-specific phonotactic patterns (Hay, Pierrehumbert, and Beckman 2004, Massaro and Cohen 1983, Pitt and McQueen 1998, McClelland and Elman 1986, Vitevich, Luce, Charles-Luce, and Kemmerer 1997, Vitevich and Luce 1999, among others). The specific claims differ among the authors, but, generally speaking, in a frequency-based account phonotactics are attributed to the listener's different sensitivity to different segment sequences; this sensitivity is established by his/her linguistic experience. Thus, if a putative phonotactic effect was observed in a pair of consonant clusters such that the perceptual bias was toward the legal cluster, one could say that such effect was obtained because the frequency of the other cluster is zero, not because it is phonotactically illegal.

In order to investigate this claim, what we need is a pair of clusters where only one is phonotactically legal and their frequencies are not compatible with the phonotactic pattern. Phonotactics and frequency are usually correlated because only legal clusters occur. However, the perceptual experiment reported in §4 will show that there is at least one case which a phonology-based account can account for but a frequency-based account cannot. Specifically, we will see that in listening to /lf/ cluster listeners invariably reported epenthetic /p/ between /l/ and /f/ rather than epenthetic /t/. We argue that a phonology-based account can explain this fact while a frequency-based account cannot, since the frequencies of /lpf/ and /ltf/ are both close to zero.

3. Phonology-based and frequency-based accounts on "phonotactic effect"

3.1. Predictions of phonology-based account

Before reporting the experiment, let us look at the predictions that phonology-based and frequency-based accounts would make about its results. As described below in detail, we compare the following six continua[4] made from word-final clusters in (3). We put silent intervals of varying durations between the sonorants and the fricatives to induce listeners to hear epenthetic stops. Silent intervals are indicated by underscores.

(3) a. /dɛlf/ – /dɛl_f/   d. /dɛmf/ – /dɛm_f/
     b. /dɛlθ/ – /dɛl_θ/   e. /dɛmθ/ – /dɛm_θ/
     c. /dɛls/ – /dɛl_s/   f. /dɛms/ – /dɛm_s/

Assuming Clements' (1987) claim, the epenthetic stops heard between the sonorants (/l/ or /m/) and the fricatives (/f/, /s/ or /θ/) are predicted to be voiceless and homorganic to the preceding sonorants. Thus, listeners are will hear /t/ after /l/ and /p/ after /m/. We now have clusters like the ones in (4):

(4) a. /dɛlf/ – /dɛl**tf**/   d. /dɛmf/ – /dɛm**pf**/
     b. /dɛlθ/ – /dɛl**tθ**/   e. /dɛmθ/ – /dɛm**pθ**/
     c. /dɛls/ – /dɛlts/   f. /dɛms/ – /dɛmps/

The bold pairs of clusters (/ltf/, /ltθ/, /mpf/, /mpθ/ and /ms/) are unattested in English (Cruttenden 2001). However, in some generative phonological theories of English syllable structure, /ltf/ is considered ungrammatical while the other four clusters are grammatical. The rationale is as follows. First, it is often claimed that English syllable codas have two segmental slots (Fudge 1968, Kiparsky 1981, Clements and Keyser 1983, Giegerich 1992, Kenstowicz 1994). When there are more than two consonants after the syllable nucleus, all of the consonants after the second one fall into the *syllable appendix*.

---

[4] Hereafter I use /lf/, /lθ/, /ls/, /mf/, /mθ/ and /ms/ to refer to each of the continua.

Given that constraint, the two consonants in the cluster /ms/ both fall into the coda with no appendix consonant. There is no apparent restriction that bans the /ms/ cluster: a voiced labial followed by a voiceless coronal (e.g. *dreamt*) is legal. The absence of this cluster is probably an accidental gap. Therefore, this cluster should be equivalent to the other attested ones in (4). There is also a widely acknowledged constraint that restricts the consonants in the syllable appendix to coronal obstruents (Blevins 1995). One way of formally representing this would be to posit a constraint of the form given in (5):

(5)   Syllable appendix constraint[5]

    * $]_\sigma$ C
         |
    [noncoronal]

The clusters /ltθ/ and /mpθ/ satisfy this constraint because /θ/ is a coronal consonant in the appendix (note that /lt/ and /mp/ are in the coda, not in the appendix), hence they are legal English clusters. The other two clusters /ltf/ and /mpf/ violate this constraint. However, as will be described below, only /ltf/ is illegal. The key idea is the *linking constraint*, which requires association lines to be interpreted as exhaustive (Hayes 1986, Itô 1986, 1989, also see Goldsmith 1990 and Itô and Mester 1993 for similar ideas). Under this idea the appendix constraint in (5) is violated if a noncoronal feature is linked exhaustively to the appendix consonant(s). Given this idea, /ltf/ violates the exhaustivity condition but /mpf/ does not, as illustrated in (6a) and (6b):

(6)   a.   mp]$_\sigma$f   b.   * lt]$_\sigma$ f
                                       | |
        [labial]   [coronal][labial]

In /mpf/, the noncoronal [labial] feature is associated to the /f/ *as well as* the /p/ consonant in the coda, whereas in /ltf/ the labial feature is linked exhaustively to the appendix consonant /f/, which violates the constraint in (5).

Since the /ltf/ cluster is phonologically illicit, we predict that listeners will perceive more /lf/ than /ltf/, ignoring the silent interval inserted between /l/ and /f/. Except for this cluster, the phonology-based account does not predict any perceptual differences among the other clusters. They are in principle all equivalent to one another.

The phonology-based account makes another prediction that is based on the markedness of labials. That is, the /m/-initial clusters will induce more epenthetic stops than the /l/-initial clusters. It is widely known that coronals show phonologically unmarked behaviors compared to labials and velars (Paradis and Prunet, 1991). Moreover, Hume, Johnson, Seo, and Tserdanelis (1999) and Jun (1995, 1996) show that this asymmetry between coronals on the one hand and labials and velars on the other is also seen in speech perception such that labials and velars are perceptually more salient than

---

[5] This constraint does not intend to explain everything about the syllable appendix consonants in English. For example, more than two consonants may be contained in the appendix, as in /glɪmpst/ 'glimpsed'.

coronals.[6] Given the assumption that /p/ is perceived in the /m/-initial clusters and /t/ in the /l/-initial clusters, it is expected that epenthetic stops are more likely to be perceived in the /m/-initial clusters than in the /l/-initial ones, because the labiality of /m/ would give listeners a stronger cue for /p/ than the coronality of /l/ would.

Based on the discussions above, we come to two predictions. One is that /lf/ will induce less epenthetic stops than the other clusters, which are predicted not to differ with one another in terms of the likelihood of perceiving epenthetic stops. The other is that /mf/, /mθ/ and /ms/ will induce more epenthetic stops than /lθ/, /ls/ and /lf/. These predictions are illustrated in (7) below:

(7)   Prediction made by phonology-based account
      /mf/ ≈ /mθ/ ≈ /ms/ > /lθ/ ≈ /ls/ > /lf/
      (ordered from high to low likelihood of perceiving epenthetic stops)

3.2.   Predictions of frequency-based account

Frequency has to do with how many times one encounters a particular lexical item or a strings of segments. In listening to ambiguous stimuli listeners are biased toward hearing frequent words or strings of segments rather than infrequent ones, because frequent items have greater strength in memory (Bybee 1998).

When considering frequency we need to distinguish two types: token frequency and type frequency. Token frequency refers to the frequency of individual items. For our current purposes, the items are the word-final consonant clusters in (4). When calculating in token frequency, each occurrence of a cluster in a corpus counts, whether it occurs in the same word or not. Type frequency is the number of items that appear in different words. All occurrences of a certain cluster in a given word count as one occurrence, regardless of how many instances of that word are observed.

Both types of frequency information are relevant in speech perception. Token frequency is important in studies that investigate the effects of so called "word frequency". It is claimed that frequent words induces greater lexical activation and more resistance to change (Bybee 1998). Type frequency is claimed to be important in accounting for the "lexical neighborhood effect", where the lexical neighborhood of an existing word or a nonsense form is defined by the set of words which differ from it by a single phoneme (Luce, Pisoni, and Goldinger 1990). Luce and his colleagues demonstrate that the lexical neighborhood density of a word has a strong influence on word perception such that it takes more time to recognize a word with high lexical neighborhood density than one with low lexical neighborhood density, because there are more "competitors" for the word with high lexical neighborhood density.[7]

---

[6] As mentioned above, the effect that Hume et al. (1999) found was small. However, it was statistically reliable, which means that the effect is not negligible.
[7] Token frequency matters in calculating neighborhood density too, as neighbors are weighted by their token frequencies.

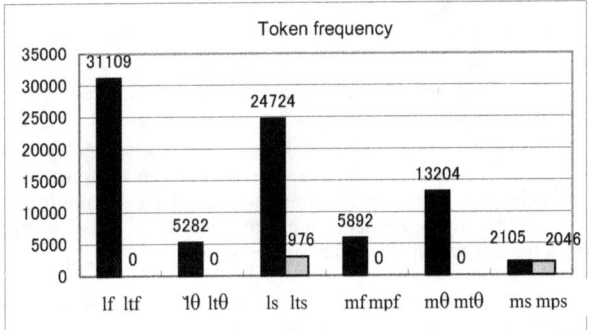

a.

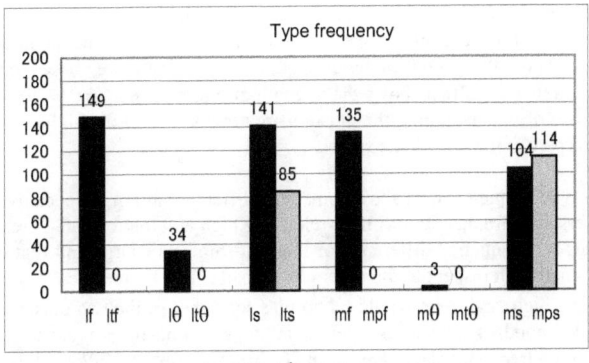

b.

Figure 1. Frequencies of the sonorant-fricative clusters and the corresponding three-consonant clusters where epenthetic stops are between the two consonants. The clusters words in the CELEX database of English (Baayen et al. 1993).

Since both token and type frequencies have been shown to be relevant in speech perception, we should consider both kinds of frequency information here. The two panels in Figure 1 show the token and type frequencies of the clusters in (4), based on the CELEX database (Baayen et al. 1993). The data are the combined frequencies of the clusters occurring within a syllable (e.g. /fɔls/ *false*) and across syllables (/maɪl.ston/ *milestone*). This is justified by the fact that epenthetic stops occur in both environments, though, as pointed out above, they occur less often across syllables (Warner and Weber 2002).

The magnitude of the frequency difference between the paired clusters with and without epenthesis would tell us how strongly listeners would be biased toward one of the clusters. That is, it is predicted that listeners would be biased toward the more frequent cluster in each pair, and that this bias would be greater when the frequency difference between the clusters is larger. To begin with, in both types of frequency data, three-consonant clusters never occur in the cluster pairs /lf/-/ltf/, /lθ/-/ltθ/, /mf/-/mpf/ and /mθ/-/mpθ/ while the two-consonant clusters in the same pairs occur at least occasionally. It is predicted from this fact that listeners will be biased toward the two-consonant clusters in these pairs rather than the corresponding three-consonant clusters, and, as a consequence, they will not hear many epenthetic stops in these environments. On the other hand, both two- and three-consonant clusters occur with frequency in the pairs in /ls/-/lts/ and /ms/-/mps/[8], which suggests that epenthetic stops are more likely to be perceived in those contexts than in those in which one of the clusters never occurs.

In the token frequency data, the /lf/-/ltf/ pair is predicted to show the strongest bias toward /lf/, since the difference between the frequency of /lf/ and that of /ltf/ is the largest among the six cluster pairs. Next strongest is the /ls/-/lts/ pair, and then the /mθ/-/mpθ/ pair. In these pairs, the bias would also be toward to the two consonant clusters. The /lθ/-/ltθ/ and /mf/-/mpf/ cluster pairs would be more or less the same, predicting weaker bias toward /lθ/ and /mf/, respectively. Finally, there would be no bias in the /ms/-/mps/ pair because their frequencies are quite close. Therefore, based on the token frequency data, the order that shows the strength of bias is /ms/ > /mf/ ≈ /lθ/ > /mθ/ > /ls/ > /lf/. For ease of understanding the clusters are arranged from the highest to the lowest likelihood of perceiving epenthetic stops.

The type frequency data show somewhat different patterns, especially for the /m/-initial clusters. The frequency of /mf/ is relatively high and that of /mθ/ is relatively low, which makes the frequency difference of the /mf/-/mpf/ pair large and that of the /mθ/-/mpθ/ pair small with respect to the other cluster pairs. The frequencies of /ms/ and /mps/ are both fairly high, and there would be no bias for either of the two clusters.[9] When we order the likelihood of perceiving epenthetic stops among the six cluster pairs on the basis of the type frequency data, they would be arranged /ms/ ≈ /mθ/ > /lθ/ ≈ /ls/ > /mf/ ≈ /lf/. Some bias against hearing epenthetic stops is predicted in all of the cluster pairs except /mθ/-/mpθ/ and /ms/-/mps/.

The two relative orderings obtained from the token and type frequency data are summarized in (8):

(8)   Predictions made by frequency-based (token and type) account
       Token: /ms/ > /mf/ ≈ /lθ/ > /mθ/ > /ls/ > /lf/
       Type: /ms/ ≈ /mθ/ > /lθ/ ≈ /ls/ > /mf/ ≈ /lf/
       (ordered from high to low likelihood of perceiving epenthetic stops)

---

[8] That the /ms/ cluster shows certain frequency comes from the fact that the data contain clusters that occur across a syllable boundary. This cluster is impossible within a syllable but possible across syllables as in /tʃɔmskɪ/ *Chomsky*.

[9] The type frequency of the /mps/ cluster is 114, which is greater than the type frequency of the /ms/ cluster by 10. I assume that this difference is negligible, and is comparable to the difference seen between /mθ/ and /mpθ/.

Compare these orders with the one obtained from the phonological account in (7), which is repeated below:

(9)   Prediction made by phonology-based account
       /mf/ ≈ /mθ/ ≈ /ms/ > /lθ/ ≈ /ls/ > /lf/
       (ordered from high to low likelihood of perceiving epenthetic stops)

We conclude this section with two remarks. First, in both accounts the /lf/ cluster is predicted to induce the fewest epenthetic stops. In both the token and type frequency data, this cluster is very frequent and the corresponding three-consonant cluster /ltf/ has zero frequency. Listeners are thus expected to often ignore the silent interval inserted as the cue for a stop. In the phonology-based account the same prediction is made, but the rationale is different: the /ltf/ cluster is ungrammatical while the corresponding /lf/ cluster is grammatical.

The second remark is that in both frequency-based accounts, more epenthetic stops are predicted to be perceived in the /m/-initial clusters than the /l/-initial clusters (though the relative order between /mf/ and /mθ/ is different between the two frequency-based accounts). Note that, as we saw in the previous section, the phonology-based account also makes this prediction. Roughly speaking, the differences between the two-consonant clusters and the corresponding three-consonant clusters are smaller for the /m/-initial clusters than the /l/-initial ones, which suggests that more epenthetic stops are expected to be perceived in the /m/-initial clusters.

4.   Experiment

4.1.   Method

4.1.1.   Participants

Seventeen listeners were recruited from an undergraduate introductory linguistics course at the University of Massachusetts, Amherst for extra credit. All were native speakers of American English and were all from the New England area. No hearing or speaking disorders were reported. Data collected from three listeners were not used for analysis because they used only two of the three possible responses.

4.1.2.   Stimuli

Six continua of CVC_1C_2 monosyllable nonwords were created by splicing CVC_1 and C_2 together with silent intervals of varying duration between them. The duration of the silent intervals varied in 10 steps from 0 ms to 100 ms, with an increment of 10 ms at each step. The CVC_1 was /dɛl/ or /dɛm/ and the C_2 was /f/, /s/ or /θ/. All possible combinations of the two types CVC_1 and three C_2 made the six experimental continua, i.e., /dɛlf/ – /dɛl_f/, /dɛlθ/ – /dɛl_θ/, /dɛls/ – /dɛl_s/, /dɛmf/ – /dɛm_f/, /dɛmθ/ – /dɛm_θ/ and /dɛms/ – /dɛm_s/ (underscores indicate silent intervals).

The stimuli were all based on tokens that were naturally produced by a male native speaker of American English. The recording took place in a sound-attenuated room. The materials were recorded onto a CD with 44.1 kHz sampling rate and 16 bit quantization level. The recorded materials were three tokens of six nonwords with no underlying stops between the sonorant consonants and the fricatives, /dɛlf/, /dɛlθ/, /dɛls/, /dɛmf/, /dɛmθ/ and /dɛms/ ('no-stop' set), and three tokens of six more nonwords where the /l/ or /m/ in each nonword was followed by a underlying homorganic voiceless stop, /dɛltf/, /dɛlts/, /dɛltθ/, /dɛmpf/, /dɛmps/ and /dɛmpθ/ ('stop' set). They were recorded in isolation without any carrier sentence, and digitized at 16 kHz/16 bits.

The CVC_1 /dɛl/ and /dɛm/ stimuli were taken from the no-stop set. For each of the C_2 /f/, /s/ and /θ/, a token that showed robust and uniform frication on spectrograms were chosen from the entire tokens in the no-stop and the stop sets. Their durations were 264ms for /ɛl/,[10] 218 ms for /ɛm/ (128 ms for /ɛ/ and 90 ms for /m/), 225 ms for /f/, 236 ms for /θ/ and 285 ms for /s/. They all had gradually falling F_0. For the /dɛl/ the F_0 value declines from 139 Hz at the initial pulse of the vowel to 112 Hz at the last pulse of the sonorant. For the /dɛm/ F_0 started at 144 Hz and ended at 129 Hz.

The reason for the decision to take /dɛl/ and /dɛm/ from the no-stop set rather than the stop set was that they inherently cautioned no acoustic indications of following stops. The acoustic analysis of the recorded materials showed that the /dɛl/ and /dɛm/ in the stop set were shorter than the ones in the no-stop set, confirming Fourakis and Port's (1986) claim. Thus, the short duration of the vocalic part would serve as a cue for the presence of a stop after it. We do not know precisely at this point how this acoustic cue could interact with the silence duration manipulated in the current experiment.[11] But it is possible that listeners would be biased toward hearing more stops if /dɛl/ or /dɛm/ were taken from the stop set, due to their short durations.[12]

A 20 ms cosine amplitude tapering was applied to the end of /dɛl/ and /dɛm/ in order to avoid abrupt offset of the sonorants, which might also help cue a stop. Also, to avoid abrupt onsets of the three fricatives which might also give listeners a cue for a stop, a 30 ms cosine amplitude tapering was applied at the beginning of the fricative.

As an example of the stimuli used in the experiment, the waveforms and the spectrograms of the stimuli /dɛl_f/, /dɛl_s/ and /dɛm_θ/ with 30 ms of silence are shown in Figure 2. In all of the spectrograms, no burst-like noise is observed in the area where silence was inserted (roughly between 270 ms and 310 ms). Also, there is no abrupt ending of the sonorant in /dɛl/ and /dɛm/, so there should be no possible acoustic cues for a stop other than the introduced silent gaps.

---

[10] /dɛl/ could not be segmented into /dɛ/ and /l/ due to the smooth transition from /ɛ/ and /l/.

[11] Some kind of trading relation would be expected between the duration of the vocalic part and silent interval (cf. Fitch et al.1980).

[12] This effect was found in a pilot experiment in which the ambiguous durations were used.

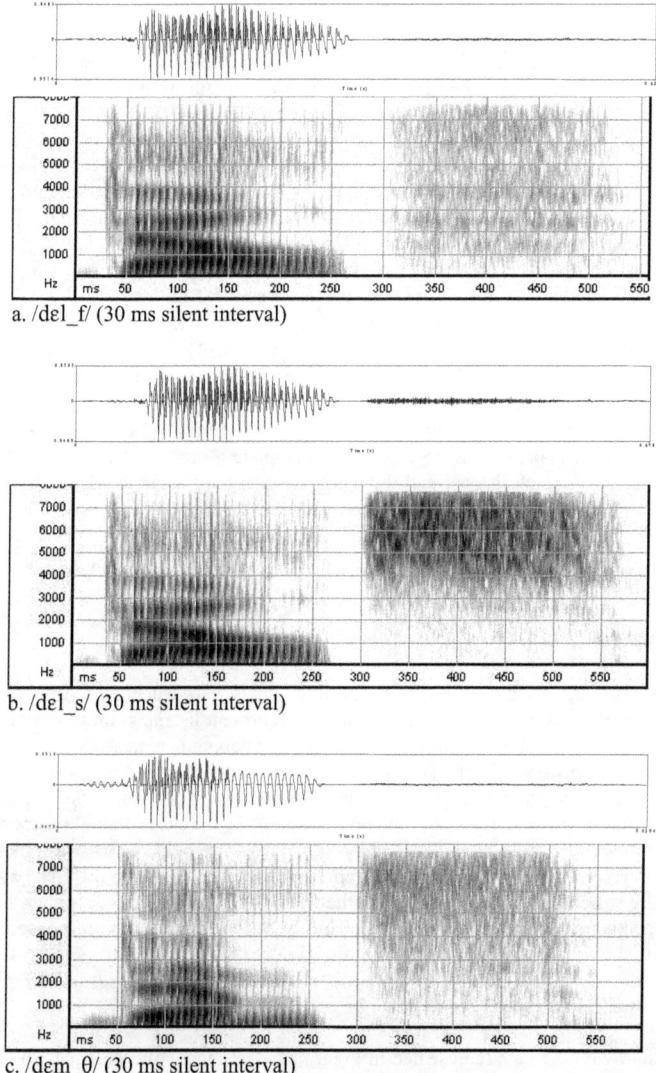

a. /dɛl_f/ (30 ms silent interval)

b. /dɛl_s/ (30 ms silent interval)

c. /dɛm_θ/ (30 ms silent interval)

Figure 2. Waveforms and spectrogram of (a) /dɛl_f/, (b) /dɛl_s/ and (c) /dɛm_θ/ with 30 ms of silence between the sonorants (/l/ and /m/) and the fricatives (/f/, /s/, and /θ/).

4.1.3. Procedure

The stimuli were transferred from the CD into an MS-DOS Windows PC, which controlled the stimulus presentation. The constructed stimuli were presented to listeners at 20 kHz sampling rate and 16 bit quantization level, one word at a time, through headphones at own comfortable volume levels. A phoneme monitoring task was adopted. The listeners were given three choices of response: /p/, /t/ or 'n(one)'. They were asked to judge whether they heard a /p/, /t/ or 'none' in a word and press corresponding buttons placed in front of them as quickly as possible. Where in the word they should focus on was not told. They had 2 seconds to give a response for each stimulus. The interstimulus interval was 2.5 seconds. The number of obtained responses per cluster type was 5 for the two end point stimuli, 10 for the two stimuli that were next to the endpoints, and 15 for the remaining stimuli. (2 endpoint stimuli × 5 responses + 2 second last endpoint stimuli × 10 responses + 7 remaining stimuli × 15 responses = 135). In each block of trials, 1 response for the 2 endpoint stimuli, 2 responses for the two second-last endpoint stimuli, and 3 responses for the remaining stimuli were obtained for each cluster type. Thus, in each block of trials, 162 stimuli were presented to the listeners ([(2×1) + (2×2) + (7×3)] × 6 = 162). It took approximately 8 minutes to complete a block of trials. Five blocks of trials were run for each listener and the whole session took about 75 minutes. The total number of stimuli each listener listened was 810 (162 stimuli × 5 blocks of trials). Before the test blocks, the listeners went through 24 practice trials, which consisted of two repetitions of the 12 endpoint stimuli of each of the six continua (2 endpoint stimuli for each continuum × 6 continua).

4.2.   Results

The response rates for the six different cluster contexts were calculated for each listener across all of the continua. The graphs in Figure 3 represent listeners' mean response rates to each continuum in the six different cluster types. The x-axis represents the 10 stimulus steps from 0 (0 ms silent interval) to 10 (100 ms silent interval) and the y-axis represents response rates. The graph in Figure 3a plots the mean /p/ response rates, the one in Figure 3b the mean /t/ response rates, and the one in Figure 3c the mean 'none' response rates. Figure 3d plots the mean rates obtained by /p/+/t/ response rates, which was obtained by "1-'none' response". Figures 4c and 4d represent the same data in a different way. It is intuitively easier to think how many epenthetic stops are perceived (4d), rather than how many epenthetic stops were ignored (4c). Thus, when we consider the response rates for "none" in the next section, we discuss Figure 3d rather than Figure 3c.

4.2.1. "Stop" vs "no stop" responses

Let us start with the data represented in Figure 3d. The most noticeable pattern is that in most of the cluster types, the listeners reported many stop percepts even when silent interval duration was short. It is striking that many stops are perceived, particularly in /lf/, /mf/ and /ms/, even at the stimulus s0, which contains no silent interval. The /lf/ cluster shows the highest stop response rates across the silent intervals. The mean response rate for /lf/ at stimulus 0 (= no silent interval) was .74. The /ls/ context shows the lowest stop response rates with most of the silent interval durations. The other contexts (/lθ/, /mf/,

/mθ/ and /ms/) fall somewhere between the two extreme clusters /lf/ and /ls/. Among the four cluster contexts, /mf/ and /ms/ showed somewhat higher response rates than /lθ/ and /mθ/, especially at shorter silent interval durations. As an overall tendency, we have found the three /m/ contexts generally elicited more stops than the /l/ contexts, except for /lf/.

Recall that the frequency data in Figure 1 make the prediction that a strong perceptual bias against hearing epenthetic stops would be observed in most of the cluster pairs, specifically at least in /lf/, /lθ/, /ls/, and /ms/. The results obtained here do not support this prediction. The /lf/ and /ms/ clusters show very high rates of stop responses at shorter silent intervals. The stop response rates for /lθ/ are also high, though not as high as for /lf/ and /ms/.

Across the silent interval steps, we can see the general tendency that as silent interval increases the stop response rates also increases. However, the magnitude of the change, i.e., the slope, varies depending on the context. The /ls/ context shows the sharpest rise across the silent interval durations, /lθ/ and /mθ/ are intermediate, and the remaining /lf/, /mf/ and /ms/ contexts show much more constant response rates across the silent intervals. We also found an asymmetry between the /l/ and /m/ series of contexts: the listeners show more drastic change in the /l/ contexts except /lf/, while they show more or less steady response rates in the /m/ contexts.

A repeated measures ANOVA was performed on the data presented in Figure 3d, with cluster type and silent interval duration as the independent variables. As our interest is in overall pattern and not in every difference between two adjacent steps, the 11 steps of the continua were reduced to three: (i) from s0 to s2, (ii) from s3 to s6, and (iii) from s7 to s10. The data for each of the three new silent interval steps were pooled and the mean was calculated for each listener. There was a significant effect of context ($F(5,65)$=7.946, $p<.004$) and silent interval ($F(2,26)$=46.449, $p<.0001$), and the interaction ($F(10,130)$=5.920, $p<.001$). Post-hoc pair-wise t-tests for contexts, with α-level corrected by the Bonferroni procedure,[13] revealed that the rates for /lf/ was significantly higher than those for /ls/ ($t(13)$=7.347, $p<.0001$) and those for /lθ/ ($t(13)$=3.906, $p<.002$), respectively, and /mf/ and /ms/ were significantly higher than /ls/, respectively (/mf/-/ls/: $t(13)$=-5.054, p<.0003; /ms/-/ls/: $t(13)$=-3.471, $p<.005$). None of the other response rates differed significantly from one another.

---

[13] Four paired samples t-tests + 3 planned contrasts (described below) were carried out. Therefore, α=.05/7=.007.

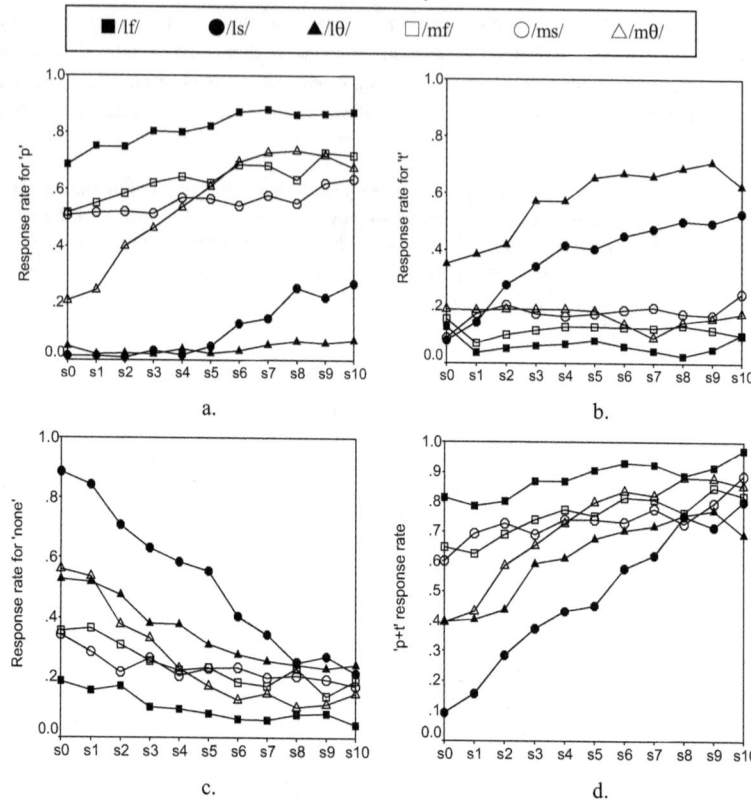

Figure 3. Mean response rates to each of the stimuli in the six different cluster contexts: (a) represents the response rates for /p/, (b) for /t/, (c) for /none/, and (d) for /p/ + /t/. Note that (c) and (d) are essentially equivalent since (d) represents the /p/+/t/ responses for ease of understanding the graph (c). On the y axis "0" indicates 0% response rate and "1" 100%. The x axis represents the stimuli, from s0 (no silent interval) to s10 (100 ms silent interval).

The reasons for the significant interaction effect in the ANOVA is the steepness difference between the response functions of the /lf/, /ms/ and /mf/ contexts, which all show more or less flat functions, and that of the /ls/ context which is much steeper. Planned contrasts showed that the rate differences between /lf/, /ms/ and /mf/ on the one hand and /ls/ on the other were significantly larger at the shortest interval than at the longest interval ($F(1,13)=24.326$, $p<.0003$). Another planned contrast comparing the differences between /lf/, /ms/, and /mf/ as one group and /lθ/ and /mθ/ as the other group was only marginally significant ($F(1,13)=7.381$, $p<.019$). Yet another planned contrast of the response rate differences between /lθ/ and /mθ/ on the one hand and /ls/ as the other was also only marginally significant ($F(1,13)=4.684$, $p<.06$). These results indicate that the /p/+/t/ response function is the gentlest in /lf/, /ms/ and /mf/, intermediate in /lθ/ and /mθ/, and the steepest in /ls/.

Based on the analyses of the /p/+/t/ response rates, we have found the following orderings for the context and the interaction effects shown in (10a) and (10b). The most notable patterns are: (i) high stop response rate for /lf/, (ii) the /m/ contexts otherwise elicited more stop responses than the /l/ contexts, and (iii) a sharper change in response rate across the silent intervals in the /l/ contexts than in the /m/ contexts.

(10)   a. Obtained context effect on /p/+/t/ response rate
          /lf/ $\approx$ /mf/ $\approx$ /ms/ > /lθ/ $\approx$ /ls/
              /mθ/
          (From high to low /p/+/t/ response rates)

        b. Obtained interaction effect on /p/+/t/ response rate
          /ls/ > /lθ/ $\approx$ /mθ/ > /lf/ $\approx$ /mf/ $\approx$ /ms/
          (From steepest to gentlest functions)

In (10a) the /mθ/ is put just below the ">" symbol, which indicates that there is no statistically significant difference between /mθ/ and the higher three contexts and the lower two contexts, respectively.

4.2.2. /p/ responses

Next, let us look at the breakdown of the 'p' response rates. They are shown in Figure 3a. One can first notice that, generally speaking, the listeners hear many /p/s in the /m/-initial contexts with the obvious exception of the /lf/ cluster case. In other words, the listeners are *p*-biased in the /m/ clusters. This result generally supports Clements' (1987) claim that an epenthetic stop shares its [place] feature with the preceding sonorant. Note that we are not in a good position to say something about the [voice] feature because listeners did not have response options of /b/ and /d/. The magnitude of the *p*-bias is relatively constant across the silent intervals, especially for /lf/, /mf/ and /ms/. That is, in the /m/ contexts the listeners kept hearing a considerable number of /p/s regardless of silent interval duration. The obvious exception is /lf/. Surprisingly, the listeners were highly *p*-biased (instead of *t*-biased) throughout the silent intervals, contradicting Clements' claim, which suggests that it is necessary to consider some other principle that governs the determination of the nature of epenthetic stops. Moreover, the degree of the *p*-bias for /lf/ was larger than the three /m/ context cases. What is especially striking is the listeners heard more /p/s in /lf/ than in /mf/, in which both the sonorant and fricative are labials while only one of the two is so in /lf/. The /mθ/ response rates for /p/ in Figure 3a showed a rise from the shorter to the longer silent intervals (especially from s0 to s7), indicating that in this context the listeners tend to ignore more silent intervals as a possible cue for a stop when they were relatively short.

The /p/ response rates were submitted to a repeated-measure ANOVA with cluster type (6 levels) and silent interval (3 levels, just as in the analysis of the /p/+/t/ responses above) as the independent variables. Both of the main effects and the interaction were significant ($F(5,65)$=55.834, p<.0001 for cluster type, $F(2,26)$=18.223, p<.0001 for silent interval, $F(10,130)$=4.888, p<.008 for the interaction;). Planned contrasts on /p/ response rate revealed that there are four distinct response patterns: (i) high flat response function (/lf/), (ii) intermediate flat function (/mf/ and /ms/), (iii) intermediate positive function (/mθ/), and (iv) low flat or slightly positive function (/ls/ and /lθ/). Specifically, the mean response rates for /lf/ were significantly greater than those averaged between /mf/ and /ms/ ($F(1,13)$=55.834, p<.0001) and those for /mθ/ ($F(1,13)$=26.325, p<.0002), respectively. There was no significant main effect on cluster type between the mean response rates for /mθ/ and those averaged between /mf/ and /ms/, but the interaction between cluster type and silent interval was significant ($F(1,13)$=5.67, p<.026). This indicates that the response function of /mθ/ is significantly steeper than /mf/ and /ms/; this is caused by the low response rates in /mθ/ at shorter silent intervals. The response rates for /mθ/ cluster and those averaged between /mf/ and /ms/, respectively, significantly differed from those averaged over /ls/ and /lθ/.

Summarizing, the responses are *p*-biased in clusters /lf/, /mf/, /ms/ and /mθ/. The p-bias is the highest in the /lf/ cluster, and /mf/, /ms/ and /mθ/ are lower. The responses for /mθ/ cluster show a sharp change depending on silent interval durations whereas those for the other clusters are more or less constant across the silent intervals.

4.2.3. /t/ responses

The mean /t/ response rates are plotted in Figure 3b. One conspicuous pattern is that the listeners heard more /t/s in the cluster environments /lθ/ and /ls/ than in the other clusters, indicating that they are more *t*-biased in /lθ/ and /ls/ than in the other clusters. These results also support Clements' (1987) claim about the featural composition of epenthetic stops. The clusters other than /lθ/ and /ls/ show consistently low /t/ response rates across the silent intervals.

Across the entire range of silent intervals, the response rates are greater for the /lθ/ cluster than the /ls/ one. The examination of the 'no-stop' responses in Figure 3c tells us that the listeners ignore silent intervals more in /ls/ than in /lθ/.

The /lθ/ and /ls/ contexts also show a similar pattern to /mθ/ in the /p/ responses in that the magnitude of the *t*-bias increases as silent interval duration increases. When the silent interval was relatively short the listeners tend ignore it but they respond to it quite reliably when the silent interval was long. Here again, we found an asymmetry between the /l/ and the /m/ contexts. That is, setting aside /mθ/ and also /lf/, the /l/-initial clusters showed a steeper response functions than the /m/-initial clusters.

A two-way repeated measure ANOVA with cluster type (6 levels) and silent interval (3 levels) as independent variables was significant for both of the main effects and interaction ($F(5,65)=12.454$, $p<.0001$ for cluster type, $F(2,26)=11.337$, $p<.0014$ for silent interval, $F(10,130)=8.275$, $p<.0002$ for interaction). Planned contrasts comparing the response rates for /lf/, those for /lθ/, and those averaged over the rest of the clusters /lf/, /mf/, /ms/ and /mθ/ revealed that /ls/ and /lθ/ were significantly different from the averaged response rates, respectively (/ls/ vs averaged rates: $F(1,13)=8.910$, $p<.011$; /lθ/ vs averaged rates: $F(1,13) = 52.158$, $p<.0001$). However, the difference between /ls/ and /lθ/ was marginal ($F(1,13) = 4.059$, $p<.065$).

4.2.4. Comparing /p/ and /t/ responses

In our discussion so far, we have found that the responses for the /lf/, /mf/, /ms/ and /mθ/ clusters are *p*-biased and those for /ls/ and /lθ/ are *t*-biased. Here, one can ask the question of whether there is any difference in the degree of bias between *p*-biased and *t*-biased cluster types. In fact, the relevant data are already provided in Figure 3d, where response rates for /p/+/t/ are represented.

The graph shows that the response rates are the lowest in /ls/ and somewhat lower in /lθ/ than in the three /m/-initial clusters and /lf/. Since we know that the /ls/ and /lθ/ responses are *t*-biased and the responses for the /m/-initial clusters and /lf/ are instead *p*-biased, it can be concluded that the degrees of the *p*-bias observed in /lf/, /mf/, /ms/ and /mθ/ are greater than those of the *t*-bias in /ls/ and /lθ/.

4.3. Discussion

4.3.1. Evaluation of the phonology-based account and frequency-based accounts

The patterns obtained in the experiment are repeated in Table 1, and are contrasted with the predictions made by the phonology- and the frequency-based accounts discussed in §3. In Table 1, a comparison of the predicted cluster orders in (a), (b) and (c) on the one hand, with the obtained results in (d) on the other reveals that the ability to account for the obtained patterns is quite poor for both frequency-based and phonology-based accounts. The result that is most damaging to the frequency-based accounts is that some clusters for which some bias against hearing epenthetic stops was predicted turned out to induce many stops. For example, in both types of the frequency-based accounts, clusters such as /lf/, /mθ/ and /ms/ are predicted to induce stop percepts less than the other clusters, since the two consonant clusters are much more frequent than the corresponding three consonant clusters /ltf/, /mpθ/ and /mts/. What we saw in the experiment is high response rates for epenthetic stops in all of these contexts.

Takahito Shinya

Prediction	Phonology-based		a.	$/mf/ \approx /m\theta/ \approx /ms/ \approx /l\theta/ \approx /ls/ > /lf/$
	Frequency- based	Token	b.	$/ms/ > /mf/ \approx /l\theta/ > /m\theta/ > /ls/ > /lf/$
		Type	c.	$/ms/ \approx /m\theta/ > /l\theta/ \approx /ls/ > /mf/ \approx /lf/$
Result	'Some stop' response		d.	$/lf/ \approx /mf/ \approx /ms/ > /l\theta/ \approx /ls/ > /m\theta/$

Table 1. Summary of the predictions made by the phonology-based and frequency-based theories and the results obtained in the experiment.

We obtained high response rates for /p/ instead of /t/ in the /lf/ cluster. However, even when we take this into consideration, the frequency accounts' ability to explain the obtained results is still poor. The /lpf/ cluster is only possible across a syllable boundary as in /hɛlp.fʊl/ *helpful*, and its frequency is far lower than that of /lf/ (/lpf/: 510 by token frequency; 3 by type frequency). Thus, a strong bias toward /lf/ is still expected (we discuss the /lf/ cluster further in the next section).

Apparently, the phonology-based account does not do any better than the frequency-based accounts. However, it does a bit better job on the /lf/ cluster, as is discussed in the next section. The prediction of the phonology-based account on clusters other than /lf/ is that no perceptual bias is expected among those clusters, which is obviously not supported by our experimental results. We found that the way a silent interval is perceived as a stop varies incredibly among those clusters. What we need is to find some other factor(s) that governs the perception of the epenthetic stops.

4.3.2. The /lf/ effect

In both phonology-based and frequency-based accounts, /lf/ was predicted to induce the fewest epenthetic stops, which is completely opposite the obtained results.

I argue that this result is problematic for the frequency-based accounts but is not anomalous in the phonology-based one. When we discussed the predictions that the two different types of account make in §3, we assumed that listeners would hear /t/s in the /lf/ cluster. However, as the results show, it was /p/ not /t/ that was perceived in this environment. The question we should ask then is whether the two different types of account are capable of distinguishing /lpf/ from /ltf/ such that /p/ is perceived more than /t/ in this context.

The phonology account can account for this asymmetry between /p/ and /t/, taking advantage of the fact that the cluster /lpf/ does not violate the syllable appendix constraint illustrated in (5) but /ltf/ does (Hayes 1986, Ito 1986, 1989). The consonant in the syllable appendix is noncoronal in /lpf/, but it shares its [labial] feature with the preceding /p/, whereby it circumvents the violation of the constraint (just as in /mpf/ in (6a)), as shown in the representation in (11a). The /ltf/ cluster, however, violates the constraint, as in (11b):

(11)   a. lp]$_\sigma$ f   b. * lt]$_\sigma$ f

[labial]   [coronal] [labial]

Because of this violation, /ltf/ was disfavored in comparison with the /lpf/ cluster. The phonology-based account therefore is compatible with the result obtained for the /lf/ context.

The frequency account, whether based on token or type frequency, cannot distinguish /lpf/ from /ltf/. We can see that /ltf/ has zero frequency both by token (Figure 1a) and type (Figure 1b). In the CELEX database, the frequency of the /lpf/ cluster is not zero but is still extremely low, since the cluster can only occur across a syllable boundary (refer to the frequency values for /lpf/ mentioned above). The only prediction the frequency-based accounts make regarding the number of epenthetic /p/s and /t/s between /l/ and /f/ is that there will be no perceptual preference (or there might be very weak preference for /p/ over /t/ since /lpf/ shows frequency while /ltf/ never occurs). Clearly, this cannot explain the strong *p*-bias observed in the /lf/ cluster.

4.3.3. Asymmetries between /l/ and /m/ contexts

One of the predictions made by the frequency-based accounts was that more epenthetic stops would be perceived in the /m/-initial clusters than in the /l/-initial ones. This prediction is generally supported by our experimental results. Aside from the special /lf/ context discussed above, we found the general tendency to be /m/ contexts > /l/ contexts: the /mf/, /ms/ and /mθ/ contexts elicited more stops than /lθ/ and /ls/ (ordering (d) in Table 1).

A phonology-based theory can also give an account of the overall "/m/ > /l/" effect by invoking the widely acknowledged idea of markedness of labials (or unmarkedness of coronals) (Paradis and Prunet, 1991). Jun (1995, 1996) and Hume et al. (1999) show that labials (and velars) are more marked than coronals, not only in formal phonological treatments but also in speech perception. They claim that labials (and velars) are perceptually more salient than coronals and thus more perceptible. Based on this claim, we can give an account of the obtained result that more labial epenthetic stops were perceived in the /m/ contexts than were epenthetic coronal stops in the /l/ contexts. The preceding /m/ is perceptually more salient, which facilitates listeners' attributing the acoustic cue for the labial sonorant to the place specification of the stop perceived in the following silent interval. Thus, the silent interval is identified as a labial stop after /m/ more often than it is identified as a coronal stop after /l/.

Since both frequency and phonology could account for the /m/ > /l/ effect, it is difficult to determine which is the right one from the data we have at hand with respect to this asymmetry. However, there is a piece of evidence for the phonological account. Warner and Weber (2001) report a similar result in Dutch, and conclude that the general perceptual bias against epenthetic /t/s and toward epenthetic /p/s and /k/s are probably not related to perceptual salience based on the phonological markedness of labials and velars. They show that their listeners are less likely to respond to /t/ than to /p/ or /k/, but they argue that that is because the epenthetic bursts in the environment of /t/ were weaker than those in the environment of epenthetic /p/ or /k/. They suggest that this may be due solely to production factors for the reasons that only one item in their materials (/np/) had an epenthetic burst but very weak, that even in an item such as /ns/ in which the epenthetic burst is widespread it is expected that epenthetic burst is weak due to the homorganicity of /n/ and /s/, and that the /ns/ stimuli itself had shorter bursts than most conditions.

This interpretation of Warner and Weber's (2001) is called into question here. The stimuli in our experiment did not contain bursts of any kind and aimed to induce stop percepts only by manipulation of the duration of the silent interval. If Warner and Weber's interpretation was correct in that only production factors were involved in the asymmetry between coronals and labials (and velars), we would obtain the result in which the listeners' bias is neutral between the two place features. This was clearly not the case in our experiment. We found that the asymmetry is still present even when there are no bursts to cue epenthetic stops.

An equally likely account that is based on the featural difference between nasals and laterals is also possible. According to this account, a stop percept is more likely after /m/ than /l/ because the former is specified [−continuant] while the latter is specified [+continuant]. [14] Listeners may have attributed the [−continuant] to the stop they perceived at the silent interval, and as a consequence more stops were perceived after /m/ than /l/.

To tease the two possibilities apart, we could use pairs of sonorant-obstruent clusters such as /ns/-/ms/ and /ns/-/ls/. The /n/ in /ns/ differs from the /m/ in /ms/ and from the /l/ in /ls/ in its place features and its manner feature, respectively. The account based on the featural contrast between [−continuant] and [+continuant] predicts equal likelihood of perceiving /t/s and /p/s for the /ns/-/ms/ cluster pair, since both /n/ and /m/ are [−continuant]. If the account based on the markedness of labial is correct, on the other hand, more epenthetic /p/s are predicted in /ms/ than epenthetic /t/s in /ns/, because /n/ is coronal while /m/ is labial. As for the /ns/-/ls/ cluster, the featural contrast account predicts more /t/ responses in /ns/ than in /ls/ while the place markedness account predicts equal likelihood of epenthetic /t/s for both clusters. I leave this issue for future research.

4.3.4. When are segments perceived?

We should ask under what environments stops, or more generally segments, are perceived. That is, why were so many epenthetic stops perceived at shorter silent intervals, or even when there was no silence, especially in /lf/ and the /m/ clusters?

The answer to the question is not obvious, especially in the environment of no silent interval. If there was no acoustic cue for a stop in the stimuli, the stop percept may be a perceptual illusion caused by some top-down process. An example of perceptual illusion is nicely demonstrated by Dupoux, Kakehi, Hirose, Pallier and Mehler (1999). They showed that Japanese listeners could not reliably judge which of a pair of nonwords such as *ebuzo* and *ebzo* they heard, while French listeners had no difficulty in discriminating them. They conclude that the phonotactic knowledge of the Japanese

---

[14] I thank John Kingston for this suggestion.

listeners corrected deviant forms in a way that they conform to the syllable structure of the language. However, it appears difficult to consider that English phonotactics is responsible for the stop percepts with no silent interval. For one thing, we saw that our phonological account permits most of the clusters used in the experiment to be grammatical. Moreover, although two consonant clusters such as /lf/ and /mf/ are perfect in English, the listeners identified very few such clusters; instead, they perceived /lpf/ and /mpf/, respectively.

The account based on the difference in feature specification for [±continuant] mentioned above could account for the many stop percepts in the /m/ clusters when there is no silent gap: listeners know that /m/ is specified for [−continuant] and /l/ for [+continuant]. The noncontinuancy of the /m/ may generate a stop percept after it. /l/ would not allow such a possibility due to its [+continuant] specification.

The high response rates for epenthetic stops in the cluster /lf/ with no silent interval, at least for now, are a mystery. However, an account is possible for the overall high stop response rates if we seek an explanation based on articulatory constraint. The /lf/ cluster consists of a lateral whose place of articulation is posterior to the following fricative. When producing this cluster, the constriction of the /f/ may precede the release of the /l/, due to an anticipatory coarticulation. When this happens, the release of the /l/ may be articulatorily masked by the labial constriction of the /f/, and therefore, no epenthetic stop would be audible. We could consider that the high response rates for /p/ in the /lf/ cluster may be a consequence of the listeners' perceptual compensation for this articulatory masking. It may be that they responded to the very short silent gaps more because they somehow knew this articulatory constraint and compensated for the masking effect of the /lf/ cluster.

4.3.5. Use of acoustic information of adjacent segments for epenthetic stops

Our experimental results generally support Clements' (1987) claim that an epenthetic stop appearing between a sonorant and an obstruent shares its place feature with the preceding sonorant. With respect to the voice feature, we cannot conclude that it is shared with the following fricative because in our experiment the listeners were not given voiced stops as response options. However, one issue we need to consider is whether they used the silent interval or the voicelessness of the fricative to identify the voice feature of the epenthetic stop. Since there is acoustically nothing during the silent interval, there is no voice bar either. There is a possibility that the listeners might have judged the absence of the voice bar in the silent portion to indicate the presence of a voiceless consonant. If this was the case, Clements' (1987) claim would be damaged, because an acoustic property which does not belong to either of the adjacent segments is used to identify the [voice] of the epenthetic stop. Clements, on the other hand, assumes that the [voice] feature does not come from the silence: it spreads from the following fricative.

Our finding that /p/ responses were dominantly perceived in the /lf/ cluster is also problematic to Clements' claim (1987) in that the epenthetic stop does not share its place feature with the preceding sonorant. Clearly, some other account is necessary. My proposal involves the markedness difference between coronals and noncoronals. First, it is reasonable to assume that when the place of articulation of a segment is to be identified where there is no robust cue, listeners use the place information of the adjacent segments. The idea is that the acoustic information regarding the place feature contained in the adjacent segments "compete" with one another to "win" the perceptual place of articulation of the silent gap. The place feature of the stop perceived in the gap is determined in accordance with the asymmetry between coronals and noncoronals. In the case of /lf/, listeners use the coronality of the /l/ and the labiality of the /f/. Since the labiality is perceptually more robust than the coronality, listeners would attribute it to the place of articulation of the stop corresponding to the silent interval. When they encounter a silent interval between /m/ and /f/, a /p/ would be perceived because [labial] is the only available feature. In /ms/ or /mθ/, one of the consonants is labial and the other is coronal, so the labial would win, and hence /p/ is the percept. Since both consonants are coronal in clusters /ls/ and /lθ/, only epenthetic /t/ is possible.

5.   Conclusion

In this paper, we have reported an experiment investigating the perception of epenthetic stops in English, with respect to how it is affected by cluster type and duration of a silent interval. We addressed three issues: (i) whether the perception of epenthetic stops is influenced by phonotactic constraints or by cluster frequency, (ii) under what circumstances listeners hear a stop, or more generally, a segment between a sonorant and a fricative, and (iii) when listeners hear an epenthetic stop, what is its featural composition and how are the acoustic properties of the adjacent segments used to identify it.

We first considered Warner and Weber's (2001) phonotactic effect seen in Dutch, which is that listeners tend to respond less to an epenthetic stop if it would form a phonotactically illegal cluster with the adjacent consonants than if it would not. However, I argued that the phonotactic effect in perception of epenthetic stops that Warner and Weber claim is invalid in that no response bias is expected between illegal two consonant clusters and the corresponding illegal three consonant clusters. Our experiment contained stimuli that predict response biases toward one of the clusters. Two different kinds of predictions were presented based on two distinct accounts – one based on frequency and one on phonological knowledge, and those predictions were tested in the experiment.

We have found that the obtained results lend little support to either the phonology-based or the frequency-based accounts. The most striking result is that an extremely high response rate for /p/ was observed in the /lf/ cluster. However, it has been shown that the phonology-based account is superior to the frequency-based account at least in one respect: it can distinguish the /ltf/ cluster from the /lpf/ cluster such that /lpf/ cluster is grammatical but the cluster /ltf/ is not, while the frequency-based account cannot distinguish these two clusters.

The experiment has shown that a silent interval plays an important role in the perception of epenthetic stops such that a long interval induces more stop percepts. The most striking result, however, is that many epenthetic stop responses were observed even when no silent interval existed. We also found that there is an asymmetry between the /m/ and /l/ clusters: more epenthetic stops are likely after /m/ than /l/. We considered two possibilities to account for this discrepancy: one based on the markedness of labial place feature and the other based on the feature [±continuant]. Which account is correct remains to be determined.

We have seen that Clements' (1987) claim that an epenthetic stop shares its place feature with the preceding sonorant and its voice feature with the following obstruent is generally supported. However, we found one case where the place feature of an epenthetic stop is shared with not the preceding sonorant but with the following obstruent, i.e., /lpf/. We proposed an account based on the phonological markedness of noncoronals, whereby the acoustic properties of the adjacent segments compete with one another to determine the place of articulation for the stop perceived between them. In the case of /lpf/, /p/ is the perceived epenthetic stop because the labiality of /f/ is perceptually more salient than the coronality of /l/.

References

Ali, L., Daniloff, R. and Hammarberg, R. (1979) "Intrusive stops in nasal-fricative clusters: An aerodynamic and acoustic investigation," *Phonetica* 36, 85-97.

Baayen, H., Piepenbrock, R., and van Rijin, H. (1993) *The CELEX Lexical Database.* Linguistic Data Consortium, University of Pennsylvania, Philadelphia.

Bybee, J. L. (1998) "The emergent lexicon," *Proceedings of the Chicago Linguistic Society: The Panels,* vol. 2, Chicago Linguistic Society 34: 421–436.

Blankenship, B. (1992) "What TIMIT can tell us about epenthesis," *UCLA Working Papers in Phonetics* 81, 17-25.

Blevins, J. (1995) "The syllable in phonological theory," In J. Goldsmith (ed.) *The Handbook of Phonological Theory.* Oxford: Blackwell, 206-244.

Booij, G. (1995) *The Phonology of Dutch.* Oxford: Oxford University Press.

Clements, G. N. (1987) "Phonological feature representation and the description of intrusive stops," *Proceedings of the Parasession on Autosegmental and Metrical Phonology,* Chicago Linguistics Society 23, 29-51.

Clements, G. N. and Keyser, S. J. (1983) *CV Phonology: A Generative Theory of the Syllable.* Cambridge, MA: MIT Press.

Cruttenden, A. (2001) *Gimson's Pronunciation of English.* Sixth edition, Edward Arnold.

Donslaar, W. van, Kuijpers, C. and Culter, A. (1999) "Facilitatory effects of vowel epenthesis on word processing in Dutch," *Journal of Memory and Language* 41, 59-77.

Dupoux, E., Kakehi, K., Hirose, Y., Pallier, C., and Mehler, J. (1999) "Epenthetic vowels in Japanese: A perceptual illusion?," *Journal of Experimental Psychology: Human Perception and Performance* 25, 1568-1578.

Fitch, H. L., Halwes, T., Erikson, D. and Liberman, A. (1980) "Perceptual equivalence of two acoustic cues for stop-consonant manner," *Perception and Psychophysics* 27, 343-350.

Fourakis, M. and Port, R. (1986) "Stop epenthesis in English," *Journal of Phonetics* 14, 197-221.

Fudge, E., C. (1968) "Syllables," *Journal of Linguistics* 4, 253-286.

Gaskell, M. G. and Marslen-Wilson, W. D. (1996) "Phonological variation and inference in lexical access," *Journal of Experimental Psychology: Human Perception and Performances* 22, 144-158.

Giegerich, H. (1992) *English Phonology: An introduction.* Cambridge: Cambridge University Press.

Goldsmith, J. (1990) *Autosegmental and metrical phonology.* Oxford: Blackwell

Hay J., Pierrehumbert, J. and Beckman, M. (2004) "Speech perception, well-formedness and the statistics of the lexicon," In J. Local, R. Ogden, R. Temple (eds.) *Phonetic Interpretation: Papers in Laboratory Phonology VI,* Cambridge: Cambridge University Press, 58-74.

Hayes, B. (1986) "Inalterability in CV phonology," *Language* 62, 321–351.

Hume, E., Johnson, K., Seo, M. and Tserdanelis, G. (1999) "A cross-linguistics study of stop place perception," In *Proceedings of the International Congress of Phonetic Sciences,* San Francisco, 2069-2072.

Itô, J. (1986) *Syllable Theory in Prosodic Phonology.* Ph.D. Dissertation, University of Massachusetts, Amherst.

Itô, J. (1989) "A prosodic theory of epenthesis," *Natural Language and Linguistic Theory* 7, 217-259.

Itô, J. and Mester, A. (1993) "Licensed segments and safe paths," *Canadian Journal of Linguistics* 38, 197-213.

Jones, D. (1966) *The Pronunciation of English.* Cambridge: Cambridge University Press.

Jun. J. (1995) *Constraint Based Analysis of Place Assimilation Typology.* Ph.D. Dissertation, UCLA.

Jun, J. (1996) "Place assimilation as the results of conflicting perceptual and articulatory constraints," *Proceedings of the West Coast Conference on Formal Linguistics* 14, 221-237.

Luce, P. Pisoni, D., and Goldinger, P. (1990) "Similarity neighborhood of spoken words," In M. Altman (ed.) *Cognitive Models of Speech Processing: Psycholinguistic and Computational Perspectives.* Cambridge, Mass: MIT Press, 122-147.

Keating, P., Blankenship, B., Byrd, D., Flemming, E., and Todaka, Y. (1992) "Phonetic analyses of the TIMIT corpus of American English," In J. J. Ohala, T. M. Nearey, B. L. Derwing, M. M. Hodge and G. E. Wiebe (eds.) *Proceedings of the International Conference on Spoken Language Processing,* 823-826.

Kenstowicz, M. (1994) *Phonology in Generative Grammar.* Cambridge, Mass: Blackwell.

Kiparsky, P. (1981) "Remarks on the metrical structure of the syllable," In W. Dresser, O. Pfeiffer, and J. Rennison (eds.) *Phonologia 1980.* Innsbruck: Innsbrucker Beitrage zur Sprachwissenschaft, 245-256.

Massaro, D. W. and Cohen, M. (1983) "Phonological context in speech perception," *Perception and Psychophysics* 34, 338-348.

Moreton, E. (2002) *Phonological grammar in speech perception.* Ph.D. dissertation, University of Massachusetts, Amherst.

Moreton, E. and Amano, S. (1999) "Phonotactics in the perception of Japanese vowel length: Evidence for long-distance dependencies," *Proceedings of the 6th European Conference on Speech Communication and Technology,* Budapest.

Ohala, J. J. (1974) "Experimental historical phonology," In J. M. Anderson and C. Ones (eds.) *Historical Linguistics II: Theory and Description in Phonology,* North-Holland Publishing Co., 353-389.

Paradis, C. and Prunet, J-F. (eds.) (1991) *Phonetics and Phonology: The Special Status of Coronals*. San Diego: Academic Press.

Pitt, M. (1998) "Phonological processes and the perception of phonotactically illegal consonant clusters," *Perception and Psychophysics* 60, 941-951.

Pitt, M. and McQueen, J. (1998) "Is compensation for coarticulation mediated by the lexicon?," *Journal of Memory and Language* 39, 347-370.

Vitevich, M. S., Luce, P. A. Charles-Luce, J. and Kemmerer, D. (1997) "Phonotactics and syllable implications for the processing of spoken nonsense words," *Language and Speech* 40, 47-62.

Vitevich, M. S. and Luce, P. A. (1999) "Probabilistic phonotactics and neighborhood activation in spoken word recognition," *Journal of Memory and Language* 40, 374-408.

Warner, N. and Weber, A. (2001) "Perception of epenthetic stops," *Journal of Phonetics* 29, 53-87.

Warner, N. and Weber, A. (2002) "Stop epenthesis at syllable boundaries," *Proceedings of the International Conference on Spoken Language Processing*, 1121-1124.

Yoo, I. W. and Blankenship, B. (2003) "Duration of epenthetic /t/ in polysyllabic American English words," *Journal of International Phonetic Association* 33, 153-164.

Department of Linguistics
South College
University of Massachusetts
Amherst, MA 01003

tshinya@linguist.umass.edu

Segmental influences on F0: Automatic or controlled?[1]

John Kingston

University of Massachusetts, Amherst

There are more things in heaven and earth, Horatio,
Than are dreamt of in your philosophy.

Hamlet, Act I, Scene 5
William Shakespeare

Aus der Kriegsschule des Lebens. Was mich nicht umbringt, macht mich stärker.
Sprüche and Pfeile 8,
Götzen-Dämmerung oder Wie man mit dem Hammer philosophirt (1889)
Friedrich Wilhelm Nietzsche

When life gives you lemons, make lemonade.
Anonymous

1. Introduction

In higher vowels and in vowels next to voiceless obstruents, the vocal folds vibrate faster than
they do in lower vowels and in vowels next to voiced obstruents. These differences are

---

[1] For comments, advice, and criticisms I am grateful to the audiences at four earlier presentations of
this paper, at: the Methods in Phonology Conference at Berkeley in May 2004, the Tone and Intonation in
Europe Conference on Santorini in September 2004, at the Phonetics and Phonology Workshop in Tokyo in
December 2004, and finally the Linguistics Department Colloquium at SUNY, Stonybrook. They have made
this a far better paper than it would otherwise have been. I retain responsibility for any errors that might have
survived these many airings.

Kathryn Flack and Shigeto Kawahara (eds.), UMOP 31, 151-183.

John Kingston

widely thought to arise as unintended and *automatic* side effects of other articulatory differences that the speaker intends to produce in pronouncing vowels differing in height or obstruents differing in voicing (Hombert, 1978; Hombert, Ohala, & Ewan, 1979; Whalen & Levitt, 1995; Connell, 2002). In this paper, I present the results of two experiments designed to test the contrary hypothesis: that these F0 differences might instead be produced by intended or *controlled* articulations (see Kingston, 1991, 1992; Kingston & Diehl, 1994 for other evidence and argument).

       Both vowel height's and obstruent voicing's effects on the rate of vocal fold vibration (henceforth *VF0* and *CF0*, respectively) have been explained as an automatic side effect of intended articulations in various ways (for VF0, see: Ohala, 1973; Ohala & Eukel, 1987; DiCristo, et al., 1979; Shadle, 1985; Steele, 1986; Silverman, 1987; Sapir, 1989; Fischer-Jorgenson, 1990; Honda & Fujimura, 1991; and for CF0, see: Halle & Stevens, 1971; Riordan, 1980; Hombert, et al., 1979; Löfqvist, Baer, McGarr, & Seider Story, 1989). However, in each case, one explanation is perhaps more plausible than the others. For VF0, raising the tongue body could pull mechanically on the vocal folds and increase their tension and rate of vibration (Ohala & Eukel, 1987). For CF0, the folds are slackened during the constriction in voiced obstruents. This slackening may be intended to aid the production of voicing itself by reducing resistance to air flow through the glottis, and thus permitting it to continue even as intraoral air pressure rises behind the oral closure downstream. Alternatively, slackening may be a result of lowering the larynx in order to sustain the transglottal pressure drop on which voicing depends by expanding the oral cavity. Larynx lowering is otherwise observed in the production of low F0 (Collier, 1974; Ewan, 1976; Honda, Hirai, Masaki, & Shimada, 1999). The slackening, whatever its cause, persists far enough into flanking vowels to slow vocal fold vibration at their edges when they occur next to voiced obstruents. There are a host of problems with both explanations, but because I have reviewed them in detail elsewhere, I won't do so again in this paper (for VF0, see: Kingston, 1991, 1992; cf. Whalen, Gick, Kumada, & Honda, 1998; Connell, 2002; for CF0, see: Kingston 1985, Kingston & Diehl, 1994). Instead I will present here a new test of the predictions of the claim that VF0 and CF0 differences are automatic side effects of other intended articulations.

       The design of this test was inspired by the insights and results of Ladd & Silverman (1984) and Steele (1986). These studies compared the occurrence and size of VF0 effects between syllables that were intonationally prominent vs those which were not, in German and English sentences, respectively. In both studies, the intonationally prominent syllable bore a high pitch accent (H*), which was realized as an F0 peak on its vowel. In the English data, syllables which were not intonationally prominent could be in the domain of a H- or a L-phrase accent, but in the German data, they were only in the domain of a L- phrase accent. Both studies obtained the same results: VF0 differences shrank or disappeared completely in syllables that weren't intonationally prominent, and in the English data, this shrinkage and disappearance occurred in H- as much as L- domains. If VF0 differences are a mechanical side effect of tongue height differences, their shrinkage and disappearance in unaccented syllables is utterly unexpected, as the vowels in the non-prominent syllables don't reduce in either language.

Even though the unaccented vowels don't reduce in either German or English, they may still be hypo-articulated compared to accented ones. Neither Ladd & Silverman nor Steele assessed differences in the pronunciation of accented vs unaccented vowels, so it's not possible to tell from their data whether VF0 differences shrank and disappeared in concert with the shrinkage and disappearance of other differences between the vowels they compared. If hypo-articulation shrinks differences tongue height, then the tongue pull hypothesis predicts that covarying F0 differences should shrink as well, in proportion to the shrinkage in tongue height differences. It seems very likely that hypo-articulation is responsible for the *some* of the shrinkage of VF0 differences in Ladd & Silverman's and Steele's data, but they shrink too much – in some instances all the way to 0 – for *all* the shrinkage to be attributed to hypo-articulation.

The dependence of VF0 differences on tone in tone languages suggests that their occurrence and size is probably not after all an automatic side effect of how hyper- or hypo-articulated the tongue height differences are. In two tone languages, Yoruba (Hombert, 1977) and Taiwanese (Zee, 1980), where words are minimally distinguished by differences in a syllable's F0, significant VF0 differences are found in high (H) but not low (L) tone syllables. Similarly, twice as many speakers produced significant VF0 differences in syllables with higher tones as did in those with lower tones in the data from four West African tone languages, Ibibio, Kunama, Mambila, and Dschang, reported by Connell (2002). None of these studies report any differences in the size of tongue height differences between vowels in H vs L tone syllables, and I'm not aware of any other studies that do, so it's unlikely that the disappearance and shrinkage of VF0 differences in L and lower tone vowels is due to any hypo-articulation of vowels bearing such tones. Taken together with the German and English results, these findings from tone languages suggest that VF0 differs between high and low vowels when they bear a H tone, although in non-tone languages, that H must be prominent and not merely high in F0.

However, in Standard Copenhagen Danish, VF0 differs significantly between high and low vowels in lexically prominent syllables, which bear a L tone, but not in immediately following syllables, whose tone is H instead (Reinholt Petersen, 1978). The easiest way to incorporate this result into an account of the effects of prosody is to distinguish between languages where (some) tones occur exclusively on prominent syllables and others appear elsewhere and languages where particular tones aren't associated with prominence. In languages of the former kind like German, English, or Danish, VF0 differs significantly only in prominent syllables, regardless of whether a H or a L tone marks that prominence, while in languages of the latter kind like Yoruba or Taiwanese, VF0 differs significantly in syllables bearing H but not L tone.

Because VF0 differences appear in some prosodic contexts but not others and the tongue height differences between vowels don't differ enough between these prosodic contexts, it's difficult or impossible to treat the VF0 differences where they do occur as an automatic side effect of another intended articulatory difference such as the height of the tongue. They might instead be plausibly treated as the product of controlled articulations in their own right.

Why would a speaker control F0 in this way? It's easy to answer this question for languages such German, English, or Danish where the VF0 differences appear in prosodically prominent contexts. Such contexts are sites where the local information content is high. In the German and English materials analyzed by Ladd & Silverman (1984) and Steele (1986), a vowel is in a prominent context because it occurs in a word bearing a pitch accent that arises in the intonation, while in the Danish materials analyzed by Reinholt Petersen (1978), the vowel occurs in the lexically prominent syllable in the word and the pitch accent is specified in the lexicon. This may be treated as a simple difference of scale: prominence and information content are high for the entire word in the German and English materials, but they are high only for one syllable in the Danish materials. Because the pitch accent is aligned with the lexically prominent syllable in the German and English words, the interval during which prominence and information content are high may actually be no more extensive in words from these languages than in the Danish words. VF0 differences are present or much larger in these prominent syllables because exaggerating the phonetic differences enhances the contrast between the vowel that occurs there and another vowel of a minimally different height that could occur there.

It is not immediately obvious why a speaker should control VF0 in tone languages such Yoruba or Taiwanese, where a H tone conveys no more prominence than a L tone. However, H tones differ from L tones in another way that motivates permitting VF0 differences at the top of the speaker's range, while limiting them at the bottom. Speakers can vary F0 far more at the top when pronouncing a H tone (Liberman & Pierrehumbert, 1984), in particular they can raise F0 more without pushing that tone's F0 target into the range of another tone. But when pronouncing a L tone at the bottom, speakers run up against a rather hard floor that prevents further F0 lowering, and they cannot raise F0 much without raising that tone's F0 target into the range of higher tone's target.

This explanation is functional, like that for the restriction of VF0 differences to prominent syllables in languages where some tones mark syllables as prominent. Both explanations refer directly to the effects the speaker is trying to have on the listener in choosing to produce VF0 differences in some contexts but not others. The two explanations nonetheless differ. VF0 differences are restricted to prominent syllables in languages where pitch accents confer phrasal or lexical prominence because they are one of the means of exaggerating contrasts in sites that the speaker estimates to have high information content. VF0 differences are restricted to H tone syllables in languages where tones convey contrasts between one word and another because they won't lead to confusion of one tone with another at the top of the speaker's range.

In this paper, I try to extend these results in two ways. The first extension is measuring the occurrence and size of VF0 differences in English syllables bearing L* as well as H* pitch accents and in H- and L- domains. This extension tests the generality of the claim that VF0 differences will only be found in prominent syllables in a language of this kind, and that these differences will be found regardless of whether the prominent syllable bears a L or a H tone. English differs from Danish in that lexically prominent syllables are only pronounced with a pitch accent when the word containing them is intonationally prominent, and F0 only differs between high and low vowels when they bear an accent (Steele, 1986). The second extension measures the occurrence and size of CF0 differences under the same prosodic manipulations. So far as I know, the effects of such manipulations on these differences have never been measured, so these results, whatever they are, will be entirely novel.

Two sets of results are reported here, pilot data collected from one speaker (the author) and a larger and more comprehensive set of data collected from four other speakers. To anticipate the results, the pilot data indicated that VF0 differences are probably controlled but CF0 are not because the presence and size of the former but not the latter vary with prosodic manipulations. These results replicate the findings of Ladd & Silverman (1984) and Steele (1986) for VF0 and show that they don't extend to CF0. The investigation of the four additional speakers turned quite differently. In the data collected from all four speakers, VF0 differed consistently across prosodic contexts, and neither its presence nor size depended on prosody. CF0 differences were small and inconsistent across speakers and contexts. These results indicate that VF0 may instead be automatic, and leaves the question open for CF0. In the discussion, I suggest that the speakers may not have treated the manipulation of pitch accents as signalling differences in local information content, so they may not have manipulated either VF0 or CF0 in ways that reflect variation in this content.

2.   Experiment 1: Pilot data

2.1.   Methods

The pilot data were collected from a single speaker, the author, who was raised in Detroit, Toronto, western Pennsylvania, northern Indiana, and Chicago.

He was recorded producing sentences containing the names, *O'Neill, O'Nail, O'Nall, O'Tail,* and *O'Dale.* F0 was measured in the middle of the high, mid, and low vowels [i], [e], and [æ] in the names *O'Neill, O'Nail,* and *O'Nall* to assess the effects of vowel height. VF0, when it differs, is expected to rank vowels [i] > [e] > [æ]. Because F0 is expected to have a particular value for the mid vowel [e], even if that value is intermediate between [i]'s high value and [æ]'s low one, [e] isn't really a control for assessing vowel height's effects on F0. F0 was also measured at the beginning and middle of the vowels following the consonants [n], [tʰ], and [d] in the names *O'Nail, O'Nall, O'Tail,* and *O'Dale* to assess the effects of consonant voicing. The sonorant [n] is not expected to raise or lower F0, so it is a genuine control context, with which the F0 values after [tʰ] or [d] may be compared. Praat's autocorrelation algorithm was used with the standard parameter settings to extract these values (Boersma & Weenink, 1992-2005; see also Boersma, 2001).

The middles of the vowels was estimated by eye from displays of the waveforms and spectrograms of the vowels. Following [n], vowel onset was the point where the second and higher formant frequencies abruptly shifted in frequency and increased in amplitude. Following [tʰ] and [d], it was the first full period of voicing after the stop release.

The names were produced in sentences of the form, *Mr. Name {sent for, has just now sent for, has just now sent impatiently for} Mr. Name by the very fastest means.* The number of words between the two names was manipulated for the purposes of another study, and the measurements reported here are collapsed across that manipulation. The sentences were pronounced with a contrastive H* pitch accent on one, the other, or neither name; when neither name was contrastively accented, the nuclear accent fell on the word "fastest". All possible combinations of names were produced except for those in which the name was the same in both positions. Twelve tokens of each name were obtained for each prosodic context (accented, other name accented, neither name accented). To determine whether F0 also differs between vowels of different heights bearing a L* pitch accent, the names *O'Neill, O'Nail,* and *O'Nall* were also pronounced in the yes/no questions corresponding to these sentences, namely *{Did, Has, Has} Mr. Name {send for, just now sent for, just now sent impatiently for} Mr. Name by the very fastest means?* The L* occurred uniformly on the name preceding the verb. Again, twelve tokens were obtained of each name.

2.2. Vowel height

Mean F0 values with 95% confidence intervals for measurements taken from the middle of the vowels in the names *O'Neill, O'Nail,* and *O'Nall* are plotted in Figure 1. The accented plotting symbols are naturally enough used in this figure for accented names; the lower case, unaccented symbols are used for the rendition in which neither name was accented; and the upper case symbols for the one in which the other name was accented. The symbols "a" and "A" represent the vowel [æ] in this figure. The values plotted on the left are for occurrences of the name before verb, and those on the right are for occurrences after the verb. The figure shows that the means for all three vowels are reliably separated from one another when the name is accented – the mean for each vowel is outside the 95% confidence interval of the adjacent vowel(s). However, when the names are unaccented, the means for all three vowels no longer differ reliably. F0 is still higher for [i] than the other two vowels in an unaccented name before the verb when the other name is accented and in an unaccented name after the verb when neither name is accented. F0 doesn't differ between unaccented [e] and [æ] under any condition.

These F0 values were used as the dependent variable in a repeated measures ANOVA in which Vowel ([i] vs [e] vs [æ]), Accent (accented, other accented, neither accented), and Position (before vs after the verb) were within-subject independent invariables. All main effects and interactions were significant, an outcome which shows that F0 values depend simultaneously on accent, position, and vowel, as can be seen in the figure. The interaction of greatest interest here is that between Vowel and Accent, where $F(4,44) = 25.183$, p < .001. In pairwise comparisons of the vowels in the different prosodic contexts, F0 is significantly higher in [i] than [e] in accented syllables ($F(1,11) = 21.116$, p = .001 and it is also significantly higher in [e] than [æ] in such syllables ($F(1,11 = 59.118$, p < .001). When neither name was accented F0 is also

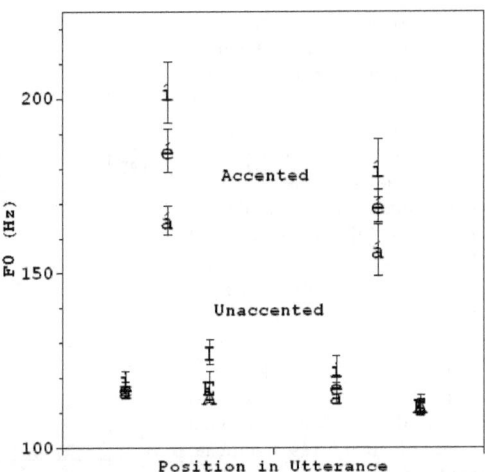

Figure 1. Mean F0 (with 95% confidence intervals) at the middle of the vowels [i] ("i" and "I"), [e] ("e" and "E"), and [æ] ("a" and "A"). Acute accent = accented vowels, upper case = other name accented, and lower case = neither name accented. The points of the left are from names preceding the verb, those on the right from those that followed it.

significantly higher in [i] than [e] ($F(1,11) = 6.707$, p < .05), but only marginally higher in [e] than [æ] ($F(1,11) = 3.470$, p = .089). None of these differences depended on the position of the name in which the measurement was taken, before vs after the verb. When the other name was accented, F0 was only significantly higher in [i] than [e] in names that preceded the verb ([i] vs [e] x Position: $F(1,11) = 20.551$, p = .001), and F0 didn't differ significantly between [e] and [æ] in either position ($F(1,11) = 2.424$, p > .10). The shrinkage and disappearance of F0 differences between vowels of different heights in unaccented syllables compared to accented ones replicates Ladd & Silverman's and Steele's findings.

Mean F0 values (with 95% CIs) are plotted in Figure 2 for the vowels in these names when they occurred bearing a L* pitch accent in yes/no questions. The means for [i] and [e] are reliably outside each other's confidence intervals, while the means for [e] and [æ] are just barely within each other's confidence intervals. The F0 values were used as the dependent variable in a repeated measures ANOVA with Vowel ([i] vs [e] vs [æ]) as the sole within-subjects independent variable. The main effect of vowel was significant (F(2,22) = 8.442, p = .002). Contrary to what is shown by the confidence intervals, planned contrasts surprisingly showed that the difference between [i] and [e]

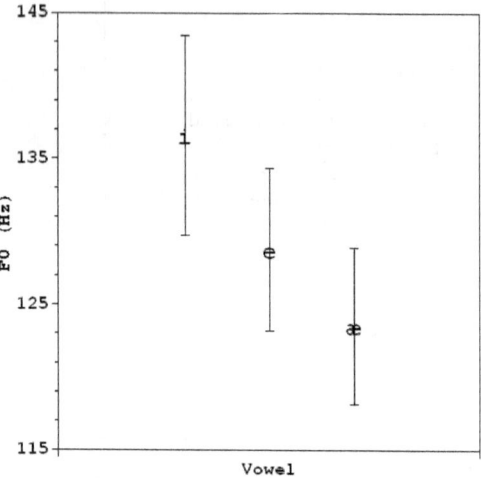

Figure 2. Mean F0 (with 95% confidence intervals) for the vowels [i], [e], and [æ] occurring in a syllable bearing a L* pitch accent.

is only marginally significant (F(1,11) = 3.987, p = .071), while that between [e] and [æ] is significant (F(1,11) = 5.428, p = .04). F0 differs less between high, mid, and low vowels when they bear L* rather than H* pitch accents, but the L* still makes the vowels prominent enough that the three heights' F0 values nearly separate reliably from one another. They certainly do so more reliably than when the vowels bear no pitch accent at all.

2.3.   Discussion

If VF0 differences were only found in H* contexts in languages such as English or only in H tone contexts in tone languages such as Yoruba, the tongue pull hypothesis would not be in jeopardy, because the larynx is pulled down when low F0 targets are produced (Ohala, 1970; Collier, 1974; Ewan, 1976; Honda, Hirai, Masaki, & Shimada,1999). Larynx lowering could counteract any upward pull on the larynx by tongue raising. However, VF0 does differ significantly in L* contexts in English (as well as Danish). VF0 also doesn't differ in whenever the tone is H, in particularly not in the context of a H- phrase accent (Steele, 1986), so the pull of the tongue doesn't always affect F0 even when larynx lowering doesn't work against it. That F0 differs between vowels contrasting for height only when they're prominent suggests that speakers produce these differences deliberately, to exaggerate segmental contrasts in sites where information content is high, rather than the tongue pulling automatically on the vocal folds.

2.4. Consonant voicing

Figure 3 displays mean F0 values at the beginning and middle of vowels following the consonants [tʰ], [d], and [n] in the names *O'Tail*, *O'Dale*, and *O'Nail* (confidence intervals have been omitted because they would have cluttered the display too much). The two measurements for each vowel are connected by a line whose quality identifies the initial consonant, solid = [tʰ], dashed = [n], and dotted = [d]. The symbol "A" identifies accented pronunciations; "U" is for unaccented pronunciations in which neither name was accented, and "O" for unaccented pronunciations when the other name was accented. The pairs of connected points on the left are for the name preceding the verb, those on the right for those following it.

F0 is substantially higher both at both the beginning and middle of the vowel following [tʰ] than [d] or [n] when that vowel is accented, and it is also higher at the beginning of the vowel following [tʰ] when the vowel is unaccented, either when the other name is accented or neither name is. F0 didn't differ between [d] and the ostensible control consonant [n] except at the beginning of accented vowels.

These F0 measurements were used as the dependent variable in a repeated measures ANOVA in which the within-subjects variables were Consonant ([tʰ] vs [d] vs [n]), Accent (name accented vs other name accentedvs neither name accented), Position (before vs after the verb), and Location of meaurement (beginning vs middle of the vowel). All main effects were highly significant, as were

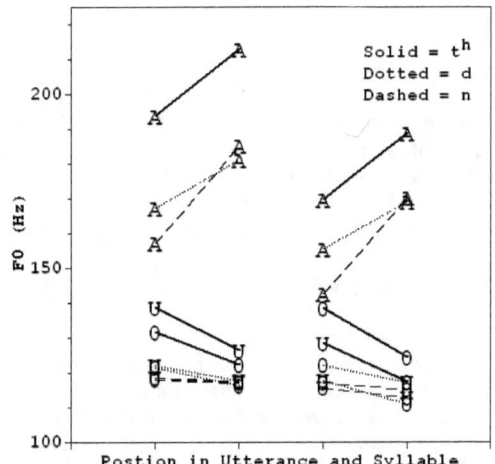

Figure 3. Mean F0 at vowel onset (left point in each pair connected by a line segment) and vowel middle (right point) following [tʰ] (solid lines), [d] (dotted lines), and [n] (dashed lines). Pairs of points on the left for names preceding the verb, those on the right for those following the verb. "A" = accented names, "O" = other name accented, and "U" neither name accented.

the interactions Location x Accent ($F(2,22) = 944.842$, $p < .001$), Consonant x Accent ($F(4,44) = 15.978$, $p < .001$), Consonant x Location ($F(2,22) = 37.616$, $p < .001$), Consonant x Accent x Position ($F(4,44) = 5.388$, $p = .001$), and Consonant x Accent x Location ($F(4,44) = 10.921$, $p < .001$).

The Location by Accent interaction reflects the rise in F0 between the beginning and middle of the vowel when the vowel is accented vs its fall between the two measurement locations when the vowel is instead unaccented. The other two-way interactions, between Consonant and Accent or Location, are superseded by the significant three-way interactions in which these variables participate, so only the latter will be interpreted. Consonant x Accent x Position is significant because the top-to-bottom order of F0 values for the name beginning with [tʰ] but not the names beginning with [d] or [n] is accented > neither name > other name when it precedes the verb, but accented > other name > neither name when it instead follows the verb. Consonant x Accent x Location is significant because in accented pronunciations F0 is higher following [tʰ] than [d] or [n] at the middle as well as the beginning of the vowel, but in unaccented pronunciations, F0 is only higher at the beginning of the vowel following [tʰ].

2.5.   Discussion

Even though the F0 differences following the three consonants are smaller in unaccented than accented words, F0 does remain reliably higher following [tʰ] than [d, n]. This outcome is quite different from that observed in the analysis of the effects of vowel height, where F0 differences between vowels of different heights effectively disappeared in unaccented syllables. Because the effects of consonant voicing are more impervious to the manipulations of prosody, they are more likely to be automatic than the effects of vowel height.

3.   Experiment 2

The principal purpose of the second experiment was to test the generality of the findings obtained in the pilot experiment by collecting similar data from more speakers. The second experiment also corrected a flaw in the materials used in the first experiment and crossed the manipulation of prosodic contexts completely with that of segment. The flaw was the choice of [n] to represent sonorants. Coarticulatory nasalization extended far enough into the following vowel to make estimates of F1 and thus vowel height unreliable. These estimates are needed to determine whether the vowels are hypo-articulated for height in non-prominent contexts and thus whether hypo-articulation is the cause of any shrinkage of VF0 differences. [n] was accordingly replaced with [l]. The full crossing of prosodic and segmental manipulations is described in the methods section that immediately follows.

3.1.   Methods

3.1.1. Participants

Two female and two male adult native speakers of English were recorded for this study. One of the female speakers was raised in Ohio (F1), the other in Toronto (F2). One of the male speakers grew up in Wisconsin (M1), the other in Northern California and Washington State (M2).

3.1.2. Materials

The effects of consonants were assessed by measuring F0 in vowels following [t], [l], and [d] in the names *Taylor*, *Layland*, and *Daylor*, while the effects of vowel height were assessed by measuring F0 in the middle of the vowels [i], [e], and [æ] in the first syllables of the names *Leland*, *Layland*, and *Lalland*. These names were produced in the place of "Name" in the following dialogues:

1. Q: What about Minnie NAME?   Who will she marry in the new year?
                            H*
    A: Minnie Name will marry Norman NAME in the new year.
            L-   H*

2. Q: What about Norman NAME?   Who will marry him in the new year?
                            H*
    A: Minnie NAME will marry Norman Name in the new year.
          H*   L-

3. Q: Will Minnie NAME marry Norman Name in the new year?
                    L*   H-
    A: No, Minnie NAME will marry Norman Name in the new year.
             H*   L-

4. Q: Will MARIE Name marry Norman Name in the new year?
             L* R   H-
    A: No, RENEE Name will marry Norman Name in the new year.
            H* F   L-

5. Q: Will Minnie Name marry LAMAR Name in the new year?
            L-   L* R
    A: No, Minnie Name will marry RENE Name in the new year.
          L-   H* F

All upper case identifies instances where the name was supposed to be pronounced in contrastive focus, i.e. with a pitch accent. Other instances of the names were supposed to be pronounced without any pitch accent aligned with the stressed syllable of the name. The intended pitch accents are given below each name, along with the expected phrase accents that specify F0 targets in unaccented pronunciations of the names. ToBI conventions are used for identifying tones (Silverman, Beckman, Pitrelli, Ostendorf, Wightman, Price, Pierrehumbert, & Hirschberg. 1992, Pitrelli, Beckman, & Hirschberg. 1994, Beckman & Ayres Elam, 1997). In dialogues 4 and 5, the first syllable of the target name immediately follows the accented syllable of the first name. As a result, the syllable where the measurements were taken occurs during the rapid F0 rise (R) from the L* pitch accent to the ensuing H- phrase accent in the questions or during the rapid F0 fall (F) from the H* to the ensuing L- in the answers.

The five names were rotated through the positions in these dialogues so that all possible combinations were produced except those in which the same name occurs twice in the same clause. There are 20 combinations in dialogues 1, 2, 4, and 5, and 15 in dialogue 3. All the combinations in each dialogue were read together in a fixed order because pilot work showed that people couldn't read these sentences in the desired way if they were randomized. All four speakers produced the dialogues twice, reading them first from 1 through 5 and then reversing the order on the second reading.

Dialogues 1 and 2 contrast in whether the accented name in the answer is early (dialogue 2) vs late. Dialogues 4 and 5 contrast similarly in whether the post-accentual rises and falls occur early (dialogue 4) or late (dialogue 5). Dialogue 3 contrasts with dialogues 1 and 2 in that the accented name in the question in dialogue 3 is expected to bear a L* pitch accent rather than the H* pitch accent it bears in the questions in dialogues 1 and 2. Finally, dialogues 4 and 5 contrast with dialogues 1-3 in having one of the names pronounced in the context of a post-accentual rise or fall, i.e. between tonal targets rather than in the context of a tonal target.

Although all four speakers pronounced these utterances with the intended accents on the intended names, they differed from one another in how they rendered the names that were supposed to be unaccented in dialogues 1 and 5. The tonal contexts in which the names were pronounced in the five dialogues are listed for each speaker in the table below:

Dialogue	M1		M2		F1		F2	
1 Q	---	a=H*	---	a=H*	---	a=H*	---	a=H*
1 A	*u=!H**	a=H*	*u=H**	a=H*	*u=H**	a=H*	*u=L-*	a=H*
2 Q	---	a=H*	---	a=H*	---	a=H*	---	a=H*
2 A	a=H*	u=L-	a=H*	u=L-	a=H*	u=L-	a=H*	u=L-
3 Q	a=L*	u=H-	a=L*	u=H-	a=L*	u=H-	a=L*	u=H-
3 A	a=H*	u=L-	a=H*	u=L-	a=H*	u=L-	a=H*	u=L-
4 Q	p=R	u=H-	p=R	u=H-	p=R	u=H-	p=R	u=H-
4 A	p=F	u=L-	p=F	u=L-	p=F	u=L-	p=F	u=L-
5 Q	u=H*	p=R	u=H*	p=R	u=H*	p=R	u=H-	p=R
5 A	*u=H**	p=F	*u=L-*	p=F	*u=H-*	p=F	*u=L-*	p=F

Table I. Tonal contexts in which the names appeared in the five dialogues in each speaker's pronunciation. The letters "a" and "u" indicate intended accented vs unaccented pronunciations, and the letter "p" indicates that the name occurs immediately after an accent ("p" for post-accentual). Transcriptions of tones follow the ToBI conventions, except that "R" and "F" indicate contexts during which F0 rises or falls between flanking tonal targets. Italicized entries pick out contexts where speakers differed from one another in their choice of tones.

3.1.3. Measurements

To assess the effects of consonants, F0 was measured at the beginning of each stressed vowel. F0 measurements used the autocorrelation algorithm in Praat (Boersma & Weenink, 2005; see also Boersma, 2001) with the standard settings for all parameters but voicing threshold, which was reduced from 0.45 to 0.2 to capture F0 values closer to voiced-voiceless transitions. For vowels following [l], this point was 5 ms after F2 began to rise steeply from its low value during the [l]; for those following [t] and [d], it was the first full glottal pulse following the stop release. The stressed vowels in all the names were followed by [l], so the end of the vowel was identified as the moment when F2 reached its minimum value in the [l]. F2 fell noticeably in all cases because the stressed vowels were all front. To assess the effects of vowels, F0 was measured across the middle 30 ms of each stressed vowel in the names beginning with [l]. The values from this 30 ms interval were then averaged, and the average was used in all further analyses of the vowels' effects.

Preliminary inspection of the data showed that for all speakers the first 5-10 periods of the vowel were often breathy voiced following [t] (see also Ní Chasaide & Gobl, 1993a,b). Because F0 is often lower in breathy than modal voice, this consonant might not raise F0 after all. Accordingly, the extent to which the voice quality was breathy was included as a possible predictor of F0 values in the analyses reported here. The breathiness measure was the difference in dB between the first and second harmonic (H1-H2), which is larger for breathier voices (Ní Chasaide & Gobl, 1997). Breathiness was measured in the same intervals that F0 was.

Finally, LPC was used with 10 coefficients to measure F1 at the middle of each vowel. The F1 measurements used in the analyses reported here are the mean calculated across the middle 30 ms of the vowel, like the F0 measurements taken from the same location. F1 was measured to assess the extent to which vowel height differences vary as a function of prosody. That is, are vowel height differences larger in accented than unaccented syllables, and if so, does the size of F0 differences between vowels of different heights correlate positively with the size of the vowel height differences as registered by F1. If F1 differences are larger and F0 differences correlate with them, then the tongue pull hypothesis would be supported.

3.2. Results

3.2.1. Analysis

So that the data from all four speakers could be combined in a single analysis, the F0 values were normalized by converting them to z-scores. The mean (μ) and standard deviation (σ) of all the F0 values from a speaker were calculated and then used to obtain the z-score corresponding to each individual value:

$$(1) \qquad z(F0) = \frac{F0 - \mu(F0)}{\sigma(F0)}$$

The values of the breathiness measure (H1-H2) and F1 were converted to z-scores in the same way.

The converted F0 values were then used as the dependent variable in linear multiple regression analyses in which the categorical independent variables (predictors) and their values were:

Predictor	Coding			
Consonant	[t] = 1	[l] = 0	[d] = -1	
Vowel	[i] = 1	[e] = 0	[æ] = -1	
Accent	Accented = 1	Unaccented = 0		
Level	High = 1	Rising = 0.5	Falling = -0.5	Low = -1
Position	Early = 1	Late = -1		

Table II. Categorical predictors and their values used in the multiple regression models.

Also included in some models were the interactions of the prosodic predictors with one another and with Consonant or Vowel. The z-transformed H1-H2 or F1 values served as a continuous predictors or dependent measures.

The values assigned to the predictors reflect expectations about how they should affect F0 values. Thus, [t] and [i] are expected to elevate F0 compared to [l] and [e], respectively, and are coded 1 compared to their 0 coding, while [d] and [æ] are expected to lower F0 compared to these intermediate sounds, and are accordingly coded -1. Because F0 is expected to be more extreme in an accented syllables, they are coded 1 while measurements taken from unaccented syllables are coded 0. This coding reflected the actual pronunciations of these words, not the intended pronunciations. As Table I above shows, all speakers pronounced words that were supposed to be accented with the intended accents, but a number of them also pronounced some words that were supposed to be unaccented with accents. Measurements taken during rising and falling intervals are also treated as unaccented for the purposes of this coding. The coding of Level as 1, 0.5, -0.5, and -1 reflects the expectation that F0 will get progressively lower from syllables in H tone contexts, down through those in Rising, Falling, and ultimately bottoming out in L tone contexts. Finally, declination will have lowered F0 little early in an utterance, so an early measurement is coded 1, while a later measurement, in a syllable that has likely undergone more declination, is coded -1.

In the first analyses, a hierarchy of five models was evaluated. After the first step in hierarchy, each subsequent model added predictors to all those in the preceding model. The hierarchy of models started with a segmental predictor, Consonant or Vowel, because F0 differences resulting from segmental contrasts are the cynosure of this paper. The breathiness measure was added next because it, too, represents a segmental property that may also influence F0. The next two models add the individual prosodic predictors, Accent, Level, and Position, and then their interactions, Accent x Level, Accent x Position, Level x Position, and Accent x Level x Position, because prosodic manipulations are expected to have very large effects on F0 that are independent of whatever segmental properties do. These independent effects must be accounted for before assessing the extent to which segmental effects depend on the prosodic context, which was done at the fifth level of the hierarchy by adding the

John Kingston

interactions between the segmental predictor, Consonant or Vowel, and the prosodic predictors, Accent, Level, and Position.

Because it will frequently be necessary to refer to the effects of interactions among prosodic predictors in determining which manipulations significantly affected F0, the values of the predictors representing these interactions are listed here (they are the products of the values of the individual predictors):

Accent x Level		Accent x Position		Level x Position		Acc x Lev x Pos Acc (Un = 0)	
Acc	Un	Acc	Un	Early	Late	Early	Late
1		1		1	-1	1	-1
(0.5)		-1	0	0.5	-0.5	(0.5)	(-0.5)
(-0.5)	0			-0.5	0.5	(-0.5)	(0.5)
-1				-1	1	-1	1

Table III. Values of predictors representing the interactions among prosodic variables. The values 0.5 and -0.5 are listed in parentheses for the Accent x Level and Accent x Level x Position interactions, because Level only takes on these values in words pronounced during a rapid rise or fall in an F0 between targets and these words are all unaccented in these materials. Acc = accented, Un = unaccented.

The values of any interactions between the segmental and prosodic predictors, including the interactions between prosodic predictors listed in Table III, will be the same as the values of the prosodic predictor when the Consonant or Vowel predictor equals 1 (when the consonant is [tʰ] or the vowel is [i]); they will be 0 when the Consonant or Vowel predictors have that value (when the consonant is [l] or the vowel is [e]), and their signs will be reversed when the Consonant or Vowel predictors equal -1 (when the consonant is [d] or the vowel is [æ]).

Tables IV and VI show the proportion of variance accounted for at each stage in the hierarchy for the analyses of the consonant and vowel data, respectively. Cross-cutting the hierarchy of models in these tables, as well as those showing the coefficients for each predictor (Tables V and VII), is a series of models that differ in which cases are included in the analysis. The hierarchy in the first column of each table includes all cases, while that in the second column excludes cases whose exclusion from the model would change the absolute value of the standardized difference in fit by more than $2*(p/n)^{\frac{1}{2}}$, where the number of parameters (p) = 16 and n = the number of cases. As n is somewhat greater than 1700, this criterion is somewhat greater than 0.19 (see the tables for precise values). These cases are outliers and excluding them naturally improves the fit of the models. The remaining columns in these tables show the effects of removing all the data from each speaker in turn. Jackknifing the analyses like this shows whether the proportions of variance and coefficient values change substantially when one or another speaker is left out. If they do, then the speakers differ from one another.

3.2.2. Consonant voicing

Table IV shows that Consonant alone accounts for very little of the variance in F0 values, from less than .01 to just over .02. The proportion of variance accounted for by Consonant is about twice as large when one of the female speakers' data is left out than when one of male speakers' data is. Adding the breathiness measure, H1-H2, changes the proportion of variance accounted for very little in any analysis except that where speaker M2 is left out, where the proportion accounted for increases nearly sixfold compared to the model in which H1-H2 isn't included. F0 is apparently far more dependent on breathiness for the other three speakers than M2. The proportion of variance accounted for jumps by an order of magnitude or more once the prosodic predictors are included. This isn't at all surprising as the tones introduced by intonation are expected to influence F0 to a much greater extent than the segments are. Adding the interactions among the prosodic predictors also boosts the proportion of variance accounted for by a good bit, adding from just under 0.12 to over 0.14 to the total. By comparison, the increment achieved by adding all the interactions between Consonant and the prosodic predictors is quite tiny, ranging from less than 0.01 to just 0.015.

Model		all (*n*=1710)	|dfit| < .19346 (1607)	- M1 (1200)	- M2 (1201)	- F1 (1220)	- F2 (1200)
1	constant, Consonant (C)	.008	.013	.007	.010	.021	.018
2	1+H1-H2	.009	.013	.013	.059	.021	.026
3	2 + Accent (A), Level (L), Position (P)	.504	.582	.563	.602	.605	.590
4	3 + AxL, AxP, LxP, AxLxP	.640	.717	.713	.724	.724	.733
5	4 + CxA, CxL, CxP, CxAxL, CxAxP, CxLxP, CxAxLxP	.649	.728	.723	.736	.733	.748

Table IV. Proportion of variance accounted for (R^2) by successive models of CF0: (1) constant and Consonant alone, (2) model 1 plus the breathiness measure (H1-H2), (3) model 2 plus the individual prosodic variables, Accent, Level, and Position, (4) model 3 plus the interactions between prosodic variables, and (5) model 4 plus the interactions between Consonant and the prosodic variables. The first column includes all cases. In the second and subsequent columns, cases are excluded whose exclusion changes the absolute value of the standardized difference in fit by more than 0.19346 (= $2*(p/n)^{\frac{1}{2}}$, where the number of parameters $p = 16$ and n=1710). In the third and subsequent columns, one speaker in turn is excluded from the model. The numbers in parentheses at the top of each column are the number of cases in the models.

In Table V are listed the values of the coefficients (βs) for each predictor obtained in model 5. Values in shaded cells aren't significant ($p > 0.10$), parenthesized values are at best marginally significant ($0.10 > p > 0.05$), and all others are significant ($p < 0.05$). No entries are given for two interactions, Accent x Level x Position and Consonant x Accent x Position for the model in which speaker F1 is left out because the values of these predictors are not independent of those of other predictors when this speaker's data is left out. The values of the coefficients for the prosodic predictors and their interactions will be discussed first because they are more consistent across models.

Both Level and Position unsurprisingly have consistently significant and large positive effects on F0, i.e. F0 is higher at the beginning of the vowel in contexts where the F0 target is higher and when the syllable occurs early in the utterance. The coefficient representing the interaction between Level and Position is also consistently significant and surprisingly negative. Given the values of this predictor (see Table III), its coefficient's negative value indicates that the higher H and R F0 targets are lowered and lower F and L targets are raised targets early in the utterance, while instead of converging, these higher and lower targets diverge late in the utterance.

The coefficients for Accent are consistently negative. It was unexpected that F0 would be lower in accented than unaccented syllables. This result probably reflects an extreme polarization of F0 targets between the L* pitch accent and H- phrase accent in the yes-no questions in dialogues 3-5, compared to the more modest differences between the H* pitch accent and L- phrase accent elsewhere. The effect of Accent nonetheless only reaches significance in models where speaker M2 is included.

The coefficient representing the interaction of Accent with Level is consistently negative. Given this predictor's values (Table III), the negative value of this coefficient indicates that F0 is lower in syllables bearing both H* and L* pitch accents compared to those bearing H- and L- phrase accents. This interpretation is consistent with the observation that the F0 differences between L* and H- targets are larger than those between H* and L- targets. In all but one model, the coefficient representing the interaction of Accent with Position is also negative. This coefficient's sign indicates that F0 is lowered in early accented syllables but raised in the late ones compared to unaccented syllables in those positions. However, in the model from which speaker F1's data were excluded, the coefficient is instead significantly positive, which shows that this speaker is responsible for the early lowering and late raising indicated by the coefficient's sign in the other models. Finally, the coefficient for the interaction between Accent, Level, and Position is consistently and significantly positive in all the models where this interaction is a predictor. In light of this predictor's values (Table III), this coefficient's sign indicates that H* and L* targets differ more from one another when they occur early than late.

Compared to these prosodic predictors, Consonant and H1-H2 have small effects. For Consonant, the effects are moreover only significant in one model, that from which speaker M2's data are excluded. In that model, the coefficient for Consonant is negative, which indicates that F0 is lower following [tʰ] compared to [n] and higher following [d] compared to [n]. These consonants were expected to affect F0 in the opposite directions. The coefficient for H1-H2 is positive in all models, which again apparently reverses the expected effect: breathier vowels having lower F0 values. However, if voice quality is more likely to be breathy following [tʰ] than the other consonants, then this coefficient's sign indicates higher F0 values following this consonant, which is the expected direction of difference.

This hypothesis is confirmed by two further analyses not reported in detail here. In the first, the roles of F0 and H1-H2 were reversed, with F0 used as a continuous predictor of H1-H2. (Only models from which outliers were excluded by the criterion described above are considered here.) In all models, the coefficients for Consonant are significantly positive: 0.926 (all speakers), 0.882 (- M1), 1.007 (- M2), 0.894 (- F1), and 0.940 (- F2), which indicates that compared to after [n], voice quality is markedly more breathy following [tʰ] and markedly less so following [d]. The coefficients for F0 in these models were also consistently positive, although not always significantly so: 0.103 (all speakers), 0.000 *ns* (-M1), 0.265 (- M2), 0.116 (- F1), and 0.032 *ns* (-F2). This finding merely confirms the positive correlation between F0 and H1-H2 observed in the models where the roles of these measures were reversed. Larger H1-H2 is more reliable at the beginning of a vowel following [tʰ] for these speakers than higher F0, but because F0 is raised more when the vowel onset is breathier, higher F0 is only somewhat less reliable.[2]

In the second analysis, H1-H2 was simply left out as a predictor of F0 values. The expectation was that the coefficients representing the effect of Consonant would turn consistently positive, as the effect that's represented by H1-H2 in the models presented in detail here is taken up by this predictor. This expectation was met, but the coefficient for this predictor was significant in only one model and marginally significant in just one more: 0.037 *marginal* (all speakers), 0.029 *ns* (- M1), 0.012 *ns* (- M2), 0.075 (- F1), and 0.031 *ns* (- F2). Categorical differences between the three consonants don't entirely capture the detailed differences in their pronunciations represented by the continuously varying H1-H2 values.

The only significant interaction between Consonant and any prosodic predictor is that with Level, whose coefficients are consistently negative in all models. Given the values of this predictor (Table III), the negative coefficient shows that F0 values in higher H and R contexts converge with those in lower F and L contexts following [tʰ] but diverge in these two contexts following [d].

---

[2] The only other significant predictors of H1-H2 values were Position, where the coefficients were significantly negative in all models and the interactions between Consonant and Position and between Consonant and Level, where for each interaction, the coefficients were significantly positive in all but one model: Consonant x Position wasn't significant for the model from which M2's data were excluded and Consonant by Level wasn't for the model from which F2's data were excluded. The negative sign of the coefficient for Position shows that vowel onsets are breathier late in the utterance than early. The positive sign of the coefficient for the interaction of Position with Consonant shows that they are breathier for an early than a late [tʰ] but breathier for a late than an early [d]. That is, the difference in breathiness between these consonants is greater early than late. Finally, the positive sign of the coefficient for the interaction of Level with Consonant is evidence that breathiness is greater following [tʰ] in the higher H and R contexts but higher for [d] in the lower F and L contexts.

Predictor	all (n=1710)	|dfit| < .19346 (1607)	- M1 (1200)	- M2 (1201)	- F1 (1220)	- F2 (1200)
constant	.173	.168	.150	.130	.196	.208
Consonant	-.033	.009	.030	-.084	(.046)	.022
H1-H2	.071	.030	.000	.095	.032	.010
Accent	-.153	(-.215)	-.243	-.178	-.790	-.218
Level	.776	.800	.784	.727	.807	.911
Position	.107	.093	.052	.110	.127	.086
AxL	-.529	-.557	-.446	-.429	-.027	-.785
AxP	-.282	-.267	-.168	-.263	.241	-.305
LxP	-.424	-.401	-.425	-.424	-.373	-.335
AxLxP	.553	.538	.450	.573		.516
CxA	.084	.090	.095	.124	(.107)	.061
CxL	-.122	-.120	-.121	-.102	-.091	-.150
CxP	-.017	-.002	.019	-.010	-.015	-.009
CxAxL	.144	.184	.146	.133	.122	.286
CxLxP	.033	.037	.029	.073	.040	.013
CxAxP	.038	-.022	-.100	-.012		.011
CxAxLxP	-.036	-.026	.036	-.062	-.003	-.063

Table V: Model coefficients for CF0: The first column includes all cases, the second and subsequent columns including only those for which absolute value of the standardized difference in fit value ≤ 0.19346 (= $2*(p/n)^{\frac{1}{2}}$, where the number of parameters (p) = 16 and n=1710), and the third and subsequent columns exclude cases for each speaker in turn (in each of these models cases are excluded whose standardized difference in fit value > 0.19346). The numbers in parentheses at the top of each column are the number of cases in the models. Cells containing non-significant coefficients are shaded, and coefficients which are marginally significant (0.05 < p < 0.10) are in parentheses.

John Kingston

3.2.3. Vowel height

Table VI shows that the proportion of variance accounted for by the model in which Vowel is the only predictor (other than the constant) is not large, but a look back at Table IV shows that this proportion is 1.5 to 2 times as large as it was for Consonant. H1-H2 was added as a predictor in the next model only to make the model hierarchy as parallel as possible to that for consonant voicing. There was no expectation that breathiness would vary with vowel height nor that it might predict F0 values in the middle of the vowel. Nonetheless, the proportions of variance accounted grew substantially in all the models once H1-H2 was added. The proportion of variance accounted for grows further between each of the remaining steps in the hierarchy in much the same way as it did in the hierarchy of models for consonant voicing effects, except that the growth between models 3 and 4, when the interactions among prosodic predictors are added is only about one half to two thirds as large for the vowel height as the consonant voicing hierarchy: increments of just 0.06-0.11 compared 0.12-0.14. The growth in the proportion of variance accounted for by adding the interactions between Vowel and the prosodic predictors is also smaller: 0.001-0.004 vs 0.01-0.015. Although this is difference is by an order of magnitude, the increments are tiny in both cases.

Model	all (n=1735)	|dfit| < .19206 (1631)	- M1 (1218)	- M2 (1223)	- F1 (1228)	- F2 (1224)
1	constant, Vowel (V)	.026	.025	.018	.024	.029	.027
2	1+H1-H2	.089	.097	.034	.387	.080	.042
3	2 + Accent (A), Level (L), Position (P)	.649	.704	.674	.762	.728	.694
4	3 + AxL, AxP, LxP, AxLxP	.728	.794	.777	.821	.806	.805
5	4 + VxA, VxL, VxP, VxAxL, VxAxP, VxLxP, VxAxLxP	.731	.796	.781	.822	.808	.808

Table VI. Proportion of variance accounted for (R^2) by successive models of VF0: (1) constant and Vowel alone, (2) model 1 plus the breathiness measure (H1-H2), (3) model 2 plus the individual prosodic variables, Accent, Level, and Position, (4) model 3 plus the interactions between prosodic variables, and (5) model 4 plus the interactions between Vowel and the prosodic variables. The first column includes all cases; in the second and subsequent columns, cases are excluded in which the absolute value of the standardized difference in fit > 0.19206 (= $2*(p/n)^{\frac{1}{2}}$, where the number of parameters $p = 16$ and $n=1735$); and in the third and subsequent columns one speaker in turn is excluded from the model. The numbers in parentheses at the top of each column are the number of cases in the models.

Even a quick glance at Table VII shows that coefficient values are more consistent across models for vowel height than they were for consonant voicing. First of all, the coefficient for Vowel is consistently significantly positive, as is that for H1-H2: F0 is higher for higher and breathier vowels. The sign and significance of the coefficients for Level and Position show that F0 is also reliably higher for vowels pronounced in higher tonal contexts and as well as being higher early in the utterance than late.

In all models, two of the interactions between prosodic predictors are significantly negative, those between Level and Accent or Position, while the coefficient for the three way interaction between these predictors is significantly positive, in all but the model from which F2's data were excluded. The signs of these coefficients match those obtained in the analysis of the consonant voicing effects and reflect the same influences of the prosodic predictors. The negative sign of the coefficient for the Level x Accent predictor shows that F0 is lower in accented syllables with H* targets and higher in those with L* targets. The negative sign of the Level x Position predictor's coefficient indicates that higher H and R targets are closer to lower F and L targets early in the utterance but farther apart late. Finally, the positive sign of the coefficient for the Level x Accent x Position predictor indicates that an accent exerts the reverse effect, pushing higher targets away from one another early in the utterance and pulling them together late.

Given the truly tiny increments in proportion of variance accounted for in going from model 4 to model 5, when the interactions between Vowel and the prosodic predictors were added, it should come as no surprise that with just one exception in one model, none of the coefficients for these interactions are significant in any model. Whatever vowel height differences do to F0 in these data, they do so independently of prosody.

Predictor	all (*n*=1735)	|dfit| < .19206 (1631)	- M1 (1200)	- M2 (1223)	- F1 (1228)	- F2 (1224)
constant	-.164	-.158	-.165	-.220	-.146	-.118
Vowel	.179	.187	.171	.210	.187	.171
H1-H2	.119	.119	.071	.255	.108	.085
Accent	-.038	-.077	-.065	-.059	-.101	-.090
Level	.964	1.001	.949	.852	1.000	1.215
Position	.043	.042	.016	.061	.064	.031
AxL	-.385	-.430	-.346	-.272	-.370	-.752
AxP	-.102	-.103	-.084	-.073	-.097	(-.112)
LxP	-.342	-.326	-.362	-.296	-.334	-.205
AxLxP	.254	.219	.214	.173	.245	.111
VxA	.053	.020	.029	-.009	.048	.022
VxL	.028	.021	.042	.004	.020	-.007
VxP	-.010	.011	.013	.030	.010	.002
VxAxL	..005	.009	.008	-.044	.016	.043
VxLxP	-.029	-.012	-.002	.015	-.011	-.037
VxAxP	.090	.059	.064	-.011	.048	.099

Table VII: Model coefficients for VF0: The first column includes all cases, the second and subsequent columns including only those for which absolute value of the standardized difference in fit value <= 0.19206 (= $2*(p/n)^{1/2}$, where the number of parameters (*p*) = 16 and *n*=1735), and the third and subsequent columns exclude cases for each speaker in turn (in each of these models cases are excluded whose standardized difference in fit value > 0.19206). The numbers in parentheses at the top of each column are the number of cases in the models.

4.   Discussion

These results are clearly at odds with expectations arising from earlier studies and the pilot data. The pilot data from a single speaker showed that F0 was reliably higher following [tʰ] than [d] regardless of prosodic context, but that F0 only differed reliably between vowels of different heights in accented syllables. This latter finding replicated those reported for

German by Ladd & Silverman (1984) and American English by Steele (1986). In this more extensive data set, from four speakers, F0 doesn't differ consistently between [tʰ] and [d], although it is higher when the vowel onset is breathier, and the vowel onset is reliably breathier when preceding consonant is [tʰ] than when it's [d]. Moreover, the robust effects of vowel height on F0 in the middle of the vowel obtained in this study are statistically independent of prosody.

The results Ladd & Silverman's and Steele's studies motivated the choice of method in this study: manipulating prosody to see whether segment's effects on F0 remained unchanged. If they did not change, then the segment's effects were to be interpreted as automatic, but if they varied systematically with the prosodic manipulation – occuring in accented = prominent syllables alone –, then they were instead to be interpreted as controlled. If the segment's effects were larger in prosodically prominent contexts, as they were in Ladd & Silverman's and Steele's results, their larger size could be interpreted as exaggeration of the contrast between that segment and a minimally contrasting one in a site where the local information content is high. This exaggeration was expected to occur when a vowel bears a pitch accent, particularly a contrastive pitch accent, because pitch accents pick out the words with high local information content.

For both vowel height and consonant voicing, these data suggest that neither vowel nor consonant pronunciations differ as function of information content. The only essential difference between the effects of vowel height vs consonant voicing is that only VF0 difference are consistently present in all contexts, while CF0 differences are more variable and also dependent on breathiness. Otherwise, by the reasoning that motivated the choice of method in this study, the results indicate that the effects of both segmental contrasts on F0 are automatic and not controlled.

What is to be done? The first thing to determine is whether these speakers pronounced the vowels and consonants differently in other ways between prominent vs not-prominent contexts. If they did not, then the prosodic manipulations were (surprisingly) unsuccessful at eliciting differences in pronunciation and other manipulations or methods must be tried.

4.1.   Vowel height

For vowel height, the F1 values at the middle of the vowel are the obvious choice for this purpose. If the articulations that implement vowel height contrasts are exaggerated, then F1 differences should be larger in prominent contexts than they are in contexts that aren't prominent. Accordingly, multiple regression analyses were run in which the dependent variable was the mean F1 spanning the middle 30 ms of each vowel. The same predictors and model hierarchy were used in this analysis as in the analyses of vowel height's effects on F0. Vowel is included in the first model in the hierarchy and this model unsurprisingly accounts for an enormous proportion of the variance, from 0.861 in the version of model 1 where F2's data are left out, to 0.877, in the version where M1's data is. The proportion of variance accounted for increases only by tiny amounts as additional predictors are added, by a total of just .009 in the version of model where F2's data are left out and by a total of just .013 in the version where M1's data is.

Table VIII is a list of the coefficients that were at least marginally significant in two or more jackknifed models. The negative coefficient for Vowel shows the expected effect of vowel height on F1: it's higher in lower than higher vowels. The positive coefficient for H1-H2 shows that breathier vowels have higher F1 values. This result is surprising given that the coefficient for H1-H2 was also positive in the analysis of vowel height's effects on F0 reported above. That analysis showed that F0 is not only higher in breathier vowels but also in higher vowels, which creates the expectation that lower vowels, i.e. those with higher F1 values, should be less not more breathy.   What needs to be recognized here that these coefficients represent those effects of H1-H2 on F0 or F1 that are independent of vowel height, and when the effects of vowel height are accounted for by another predictor, Vowel, both F0 and F1 increase with breathiness.

F1 is also higher in accented than unaccented syllables, suggesting that the mouth is opened wider in accented vowels (Harrington, Fletcher, & Beckman, 2000; Erickson, 2002), but this effect is significant in only one model and marginally significant in one other. The coefficients representing the interactions of Level with Accent or Position are both negative but significantly so in only one jackknifed model for Level x Accent and in only two for Level by Position. Given the values of the Level by Accent predictor (Table III), the sign of this coefficient indicates that F1 is lower, i.e. the vowel is higher when F0 is higher. The negative coefficient for the Level x Position predictor indicates that F1 values in syllables in higher H and R vs lower F and L contexts are pulled together early in the utterance but they are pushed apart late.

The coefficients representing the interaction between Vowel and Accent are more consistently significant. They are also uniformly negative, which means that an accented high vowel has a lower F1 value and an accented low vowel has a higher F1 than these vowels would have if they were unaccented. That is, height differences *are* exaggerated in accented syllables compared to unaccented ones. Yet F0 differences are no greater between high and low vowels when accented than unaccented – the Vowel x Accent interaction wasn't even marginally significant in the analysis of vowel height's effect on F0. The presence of an accent does alter how much vowels differ in tongue height, but despite doing so its presence doesn't alter how much they differ in F0.

Predictor	all (*n*=1735)	|dfit| < .19206 (1631)	- M1 (1200)	- M2 (1223)	- F1 (1228)	- F2 (1224)
constant	.110	.113	.100	.143	.089	.110
Vowel	-1.249	-1.278	-1.264	-1.284	-1.306	-1.267
H1-H2	.028	.055	.036	.093	.048	.052
Accent	.098	.204	.223	.155	.064	(.201)
A x L	-.109	(-.191)	-.212	(-.182)	.007	-.182
L x P	-.029	-.035	-.042	-.034	-.052	-.008
V x A	-.322	-.266	(.046)	-.264	-.277	-.265

Table VIII: Model coefficients for F1: The first column includes all cases. The second and subsequent columns include only those for which absolute value of the standardized difference in fit value <= 0.19206 (= $2*(p/n)^{\frac{1}{2}}$, where the number of parameters p = 16 and n=1735). The third and subsequent columns exclude cases for each speaker in turn (in each of these models cases are excluded whose standardized difference in fit value > 0.19206). The numbers in parentheses at the top of each column are the number of cases in the models.

This result does therefore support the hypothesis that F0 differences between vowels of different heights are automatic. However, there is a problem. The most plausible explanation of these differences has been that the raising the tongue somehow stretches the folds, while lowering it somehow slackens them. This explanation is supported by Ohala & Eukel's (1987) finding that F0 differences between vowels of different heights were larger when the jaw was propped open by a bite block and speakers had to raise the tongue more independently of the jaw to get it close enough to the palate to successfully produce a high vowel. The problem is this: the presence of an accent has just been shown to exaggerate F1 differences between high and low vowels, and this exaggeration is presumably achieved by raising the tongue more in high vowels and lowering it more in low vowels, yet despite these larger differences in tongue height, F0 differences are no greater in accented than unaccented vowels.

Why doesn't raising and lowering the tongue more also raise and lower F0 more if F0 differences between vowels are a byproduct of the tongue pulling on the vocal folds? The answer may be that the larger F1 differences are a byproduct of raising or lowering the *jaw* more in accented than unaccented syllables, and that the tongue itself is not raised or lowered more independently of the jaw than it is in unaccented syllables. Harrington, et al. (2000) and Erickson (2002) both show that the effect of an accent is to increase jaw movement and that the tongue actually moves independently of the jaw to ensure that the desired vowel target is still reached. So the finding that the size of F1 but not F0 differences varies with the presence of an accent may not be such a serious problem after all.

4.2.   Consonant voicing

For consonant voicing, voice onset time (VOT) might differ more in prominent than non-prominent contexts. Z-transformed VOT values from the four speakers were used as the dependent variable in the same model hierarchy as was used to analyze the effects of consonant voicing on F0 at vowel onset, except that the Consonant predictor had only two values, 1 for [tʰ] and -1 for [d], because no VOT measurements can be made on [l].

Unsurprisingly, the first model in the hierarchy, whose only predictor is Consonant, accounts for a very large proportion of the variance in VOT, ranging from .801 for the model which includes all the cases to 0.891 for the model that excludes data from speaker M1. The increments in the variance accounted from adding all the other predictors are modest, ranging from just 0.019 to 0.033. Again the model including all the cases accounts for the smallest proportion of the variance at step 5 in the hierarchy, 0.839, and the one excluding speaker M1 accounts for the most at this step, 0.910.

Table IX lists the coefficients that were at least marginally significant in two or more of the jackknifed models. The results are strikingly uniform and straightforward. Naturally, the coefficient for Consonant is positive as VOT is longer in [tʰ] than [d]. The coefficient for H1-H2 is consistently negative, which indicates that VOT is shorter when following vowel onsets are breathier. This tradeoff is not surprising because VOT and breathiness are both correlates of glottal spreading. Tokens differ from one another in whether voicing begins while the glottis is still appreciably spread, in which case the vowel will begin with a breathier voice quality and VOT will be shorter, or if voicing doesn't begin until the glottis is no longer particularly spread, in which case the vowel will begin with a more modal, less breathy voice quality following a longer VOT. The positive coefficient for Accent shows that VOT values are longer in accented than unaccented syllables.   The coefficient for Level is consistently negative, although only significantly so in two of the jackknifed models. The sign of this coefficient shows that VOT values are shorter in H or R than F or L contexts. The positive coefficient for Position indicates that VOT values are longer early in the utterance than late. Level and Position interact significantly and the coefficient for this interaction is positive. VOT values are thus longer in H and R than F or L contexts early in the utterance, but shorter late. Finally, a single interaction between the segmental and prosodic predictors is significant, Consonant x Position, and its coefficient is consistently positive. VOT values differ more between [tʰ] and [d] early in the utterances than late.

The significance of the coefficient for Accent doesn't indicate that speakers exaggerate VOT differences between [tʰ] and [d] in accented sylllables compared to unaccented ones. They merely make them longer. Exaggeration of differences would only be indicated by a significantly positive coefficient for the interaction between Consonant and Accent. Though uniformly positive, this interaction's coefficient was at best marginally significant and that in just one model, the one including all the cases. VOT differences between [tʰ] and [d] are only consistently exaggerated early in the utterance compared to late, as shown by the significantly positive coefficient for the Consonant by Position interaction. An early position in the utterance would often, perhaps even typically, have higher information content than a later position since content is less predictable early, but an early position is not one of greater prominence like an accented syllable. These speakers don't exaggerate VOT differences between [tʰ] and [d] in prominent syllables any more than they do F0 differences in the vowels following these consonants.

Predictor	all (*n*=1133)	|dfit| < .23767 (1067)	- M1 (812)	- M2 (788)	- F1 (805)	- F2 (796)
constant	-.131	-.117	-.111	-.098	-.117	-.129
Consonant	.867	.877	.892	.913	.889	.849
H1-H2	-.070	-.066	-.061	-.109	-.059	-.062
Accent	.220	.180	.171	.177	.285	.193
Level	-.043	-.033	-.043	-.025	-.016	-.044
Position	.045	.051	.036	.045	.064	.057
L x P	.047	.058	.057	.051	.056	.067
C x P	.049	.054	.036	.053	.064	.068

Table IX: Model coefficients for VOT: The first column includes all cases. The second and subsequent columns include only those for which absolute value of the standardized difference in fit value <= 0.23767 (= $2*(p/n)^{1/2}$), where the number of parameters p = 16 and n=1133). The third and subsequent columns exclude cases for each speaker in turn (in each of these models cases are excluded whose standardized difference in fit value > 0.23767). The numbers in parentheses at the top of each column are the number of cases in the models.

4.3.   Summary and concluding remarks

Taken together, these findings show that prominence doesn't prompt speakers to increase the distinctiveness of the [voice] contrast between these two consonants, at least as measured by VOT and F0. These findings also dovetail with the conclusions drawn from the further investigation of VF0 differences, which also weren't exaggerated in prominent syllables, although another correlate of vowel height contrasts, F1, was. In short, unlike Ladd & Silverman (1984) or Steele (1986), I was unable to get manipulations of prosody to alter articulations of segments in ways that would affect F0, for either vowel height or consonant voicing.

Again, the question can be asked: what is to be done? The answer, I think, is to try manipulating prosody such that one can be confident that the presence or absence of a pitch accent does correlate with the waxing and waning of local information content. The materials used in this study were designed to produce such a correlation, and the speakers certainly hyper-articulated accented words in these utterances and hypo-articulated unaccented ones. However, those differences in the extent of hyper- vs hypo-articulation may not have reflected variation in local information content because each sentence was simply a permutation of the names. New materials are being designed to correct this error.

Even if this error can be corrected, the present results remain a problem for interpreting Ladd & Silverman's and Steele's findings, as they showed that VF0 differences apparently depended solely on the presence of a pitch accent – as did the pilot results reported here. The speakers in experiment 2 certainly produced some words with H* and L* pitch accents and just as certainly produced others without any such accents, yet VF0 differed as much in the unaccented as the accented syllables. This finding suggests that these differences are not after all dependent on a pitch accent. Instead, in anticipation of the results of the new manipulations of prosody, we might conclude that these differences and the presence of a pitch accent are independently controlled means of conveying local information content.

References

Beckman, Mary E., and Gayle Ayers Elam. 1997. Guidelines for ToBI labelling. (version 3.0). Ms., The Ohio State University.

Boersma, Paul 2001. Praat, a system for doing phonetics by computer. Glot International 5:9/10: 341-345.

Boersma, Paul, and David Weenink 2005. Praat: doing phonetics by computer (Version 4.3.01) [Computer program]. Retrieved from http://www.praat.org/.

Collier, René. 1975. Physiological correlates of intonation patterns. Journal of the Acoustical Society of America 58:249-55.

Connell, Bruce. 2002. Tone languages and the universality of instrinsic F0: Evidence from Africa. Journal of Phonetics 30: 101-129.

DiCristo, A., D. J. Hirst, and Y. Nishinuma. 1979. L'estimation de la F0 intrinsèque des voyelles: Etude comparative. Travaux de L'Institut de Phonétique D'Aix-en-Provence 6: 149-176.

Erickson, Donna. 2002. Articulation of extreme formant patterns for emphasized vowels. Phonetica 59:134-149.

Ewan, William G. 1976. Laryngeal behavior in speech. Ph.D. dissertation. University of California, Berkeley.

Fischer-Jørgenson, Eli. 1990. Intrinsic F0 in tense and lax vowels with special reference to German. Phonetica 47: 99-140.

Halle, Morris, & Kenneth N. Stevens. 1971. A note on laryngeal features. Quarterly Progress Report, Research Laboratory in Electronics 101: 198-213. Cambridge, MA: Massachusetts Institute of Technology.

Harrington, Jonathan, Janet Fletcher, and Mary E. Beckman. 2000. Manner and place conflicts in the articulation of accents in Australian English. In Papers in laboratory phonology V: Acquisition and the lexicon, ed. Michael Broe and Janet B. Pierrehumbert, 40-51. Cambridge, UK: Cambridge University Press.

Hombert, Jean-Marie. 1977. Consonant types, vowel height, and tone in Yoruba. Studies in African Linguistics 8: 173-190.

Hombert, Jean-Marie. 1978. Consonant types, vowel quality, and tone. In Tone: A linguistic survey, ed. Victoria Fromkin, 77-111. New York: Academic Press.

Hombert, Jean-Marie, John J. Ohala, & William G. Ewan, 1979. Phonetic explanations for the development of tones. Language 55: 37-58.

Honda Kiyoshi, & Osamu Fujimura. 1991. Intrinsic vowel F0 and phrase-final lowering: Phonological vs biological explanations. In Vocal fold physiology: Acoustic, perceptual, and physiological aspects of voice mechanisms. ed. J. Gauffin and B. Hammerberg, 149-157. San Diego: Singular Publishing Group.

Honda, Kiyoshi, H. Hirai, S. Masaki, and Y. Shimada. 1999. Role of vertical larynx movement and cervical lordosis in F0 control. Language and Speech 42: 401-411.

Kingston, John. 1985. The phonetics and phonology of the timing of oral and glottal events. Ph.D. dissertation, University of California, Berkeley.

Kingston, John. 1991. Integrating articulations in the perception of vowel height. Phonetica 47: 149-179.

Kingston, John. 1992. The phonetics and phonology of perceptually motivated articulatory coordination. Language & Speech. Festschrift for John J. Ohala 35: 99-113.

Kingston, John, & Randy L. Diehl. 1994. Phonetic knowledge, Language 70: 419-454.

Ladd, D. Robert, & Kim E. A. Silverman. 1984. Vowel intrinsic pitch in connected speech. Phonetica 41: 31-40.

Liberman, Mark Y., and Janet B. Pierrehumbert. 1984. Intonational invariance under changes in pitch range and length. Language sound structure, ed. Mark Aronoff and Richard Oehrle, 157-233. Cambridge, MA: MIT Press.

Löfqvist, Anders, Thomas Baer, N. McGarr, & R. Seider Story. 1989. The cricothyroid muscle in voicing control. Journal of the Acoustical Society of America 85: 1314-1321.

Ní Chasaide, Ailbhe, & Christer Gobl. 1993a. Dynamic variation of the voice source in VCV sequences: Intrinsic characteristics of selected consonants. Proceedings of the first review meeting of the Esprit Basic Research Action, No. 6975: SPEECH MAPS, 2, 44. Grenoble: Institut de la Communication Parlée.

Ní Chasaide, Ailbhe, & Christer Gobl. 1993b. Contextual variation of the vowel voice source as a function of adjacent consonants. Language and Speech 36: 303-330.

Ní Chasaide, Ailbhe, & Christer Gobl. 1997. Voice source variation. In The handbook of the phonetic sciences, ed. W. J. Hardcastle & John Laver, 427-461. Oxford, UK: Blackwells.

Ohala, John J. 1970. Control and production of speech. University of California, Los Angeles, Working papers in phonetics 15.

Ohala, John J. 1973. The physiology of tone. In Consonant types and tone, Southern California Occasional Papers in Linguistics, 1 ed. Larry M. Hyman, 1-14.

Ohala, John J., & Brian Eukel. 1987. Explaining the intrinsic pitch of vowels. In In honour of Ilse Lehiste, ed. R Channon and L. Shockey, 207-215. Dordrecht: Foris.

Pitrelli, John F., Mary E. Beckman, and Julia Hirschberg. 1994. Evaluation of prosodic transcription labeling reliability in the ToBI framework. Proceedings of the 1994 International Conference on Spoken Language Processing, v. 1, 123-126. Yokohama, Japan.

Reinholt Petersen, N. 1978. Intrinsic fundamental frequency of Danish vowels. Journal of Phonetics 6: 177-189.

Riordan, Carol J. 1980. Larynx height during English stop consonants. Journal of Phonetics 8: 353-360.

Sapir, S. 1989. The intrinsic pitch of vowels: Theoretical, physiological, and clinical considerations. Journal of Voice 3, 44-51.

Shadle, Christine. 1985. Intrinsic fundamental frequency of vowels in sentence context. Journal of the Acoustical Society of America 78: 1562-1567.

Silverman, Kim E. A. 1987. The structure and processing of fundamental frequency contours. Ph.D. dissertation. Cambridge University.

Silverman, Kim E. A., Mary E. Beckman, John F. Pitrelli, Mari Ostendorf, Colin Wightman, Patti J. Price, Janet B. Pierrehumbert, and Julia Hirschberg. 1992. TOBI: A standard for labeling English prosody. Proceedings of the 1992 International Conference on Spoken Language Processing, Vol. 2, 867-870, Banff, Canada.

Steele, Shirley A. 1986. Interaction of vowel F0 and prosody. Phonetica 43: 92-105.

Whalen, Douglas H. Andrea G. Levitt. 1995. The universality of intrinsic F0 of vowels. Journal of Phonetics 23: 349-366.

Whalen, Douglas H., Bryan Gick, M. Kumada, and Kiyoshi Honda, 1999. Cricothyroid activity in high and low vowels: Exploring the automaticity of intrinsic F0. Journal of Phonetics 27: 125-142.

Zee, Eric. 1980. Tone and vowel quality. Journal of Phonetics 8: 247-258.

Linguistics Department
South College
University of Massachusetts, Amherst
Amherst, MA 01003

jkingston@linguist.umass.edu